GEOGRAPHY *and the Human Spirit*

GEOGRAPHY *and the Human Spirit*

Anne Buttimer

With a Foreword by Yi-Fu Tuan

The
Johns Hopkins
University Press

Baltimore

London

Published in cooperation with the Center for American Places, Harrisonburg, Virginia.

Printed in the United States of America on acid-free paper

The Johns Hopkins University Press
2715 North Charles Street
Baltimore, Maryland 21218–4319
The Johns Hopkins Press Ltd., London

Library of Congress
Cataloging-in-Publication Data

Buttimer, Anne.
Geography and the human spirit / Anne Buttimer; with a foreword by Yi-Fu Tuan.
p. cm.
Includes bibliographical references and index.
ISBN 0-8018-4338-3 (alk. paper)
1. Human geography—Philosophy.
I. Title.
GF21.B87 1993
304.2—dc20 92-25972

A catalog record for this book is available from the British Library.

Contents

Illustrations

Foreword

MILLIONS OF PEOPLE have seen the picture of the planet earth shot from the moon. By now, we may even feel nonchalant about it. Yet, when the picture first appeared, many were struck by the beauty of the earth, our only home, which presents itself to us as a marbled globe, wreathed in bands of white, green, and blue, a fertile island hospitable to living things in a sterile black sky. "This little world . . . this precious stone set in the silver sea, this blessed plot," we suddenly realize, applies not only to England (as Shakespeare would have it) but to the planet itself.

Geography is the study of the earth as the home of human beings. Home. How varied and resonant are the meanings of that word. Home is the envelope of air; it is the continents and the oceans, the deserts and the forests. Home is the humanly modified worlds of farms and gardens, towns and cities. Home is parish, province, nation-state, Asia and Africa, the North Atlantic Alliance and the Third World. Home is sociality—it is types of human connectedness at all scales. These range from intimate exchanges in family and neighborhood to communication across thousands of miles via electronic media, those myriads of invisible lines that encircle the earth, creating an extra sheath of shared thought and feeling. Last, but certainly not least, home is the mutual dependence and sociality of all living things—plants, animals, and people.

Understanding the earth as our home—understanding the meaning of the verb *to dwell*—is an immense challenge to the human spirit. Scholars have tried to document the challenge and our response by writing histories and philosophies of geography. Sometimes they focus narrowly, restricting themselves to a period such as the Middle Ages or to a place such as France, or broadly in regard to period but narrowly in the sense of confining the cast of characters to geographers. Or perhaps both the time period and the stage may be broad, as is also the definition of "geographical thinker," but the ideas examined may be deliberately confined to a small set. The exemplary work of this last genre is, of course, Clarence Glacken's survey of geography and the human spirit. But is it possible to be more inclusive? Can one take the whole of Western history and include not only geographers but also philosophers who have influenced geography? Can one situate geographical ideas in the context of not only intellectual but also sociopolitical climate, thus linking societal needs and institutions to the geographical enterprise? Is it possible to see geography as having a core of ideas which have nevertheless undergone so many changes that their moral if not intellectual significance is drastically altered—even reversed, such that, for instance, "organic whole" as applied to "interdependence within a region" becomes the monster of *Lebensraum* and imperialism?

Yes, I would have said. It is possible but not probable—until, that is, I read Anne Buttimer's book, which shows not only that all this can be done but that it can be done within short compass. The technique, which requires inspiration, is to build principal theses and themes, personalities and historical events, on key sets of firm yet resilient metaphors. These include Phoenix, Faust, Narcissus; mosaic, organic whole, mechanism, arena, and the protean meanings of humanism.

Let me briefly consider the term *humanism,* following Anne's lead but giving the variant meanings twists and extensions of my own. *Humanus,* as opposed to *barbarus,* denotes high cultural and intellectual achievements. Anne puts on the Roman cloak of *humanus* when she draws our attention to human excellence—to the great geographical concepts as well as the ambitious attempts to transform nature that took place in classical antiquity, the Renaissance, the Enlightenment, and the modern period. Humanism, however, also implies universalism: "I am a man, I regard nothing that is human as alien to me" (Terentius). Anne's own aspiration is universalist. Although the focus of her book is Western, she is very much aware of the contributions of other cultures and civilizations to the understanding and development of the earth as human habitat. Humanism, in the Heideggerian sense, is a patient listening to reality, letting it reveal itself rather than forcing it to yield its truth, essence, or value by Faustian means. Anne presents the phenomenological tradition in geography with warm sympathy. Humanism has frequently been linked with *paideia,* or education. In Europe, though not in America, geography has been an important school subject since the nineteenth century. In histories of geography, however, it is rare that the subject's pedagogic thrust and (beyond the schoolroom) its role in the general enlightenment of the public have received any serious attention. Anne's study is an exception: for her, *paideia* is a key word. Humanitarianism, as a consciously held ideal, is rooted in humanism. Anne explores the applications of geography, not only in the direction of efficiency (better bus routes and sites for shopping malls, etc.), but also in the direction of a better life—a more humane life. And then there is the challenge of a fundamental value in humanism—Socrates' "Know thyself," which, as Anne shows, is in danger of becoming in our time, "Doubt thyself," and, ungenerously, "Doubt the other—and everything other people think or do." Postmodernism and its twin, cultural relativism, are a form of liberation, a Phoenix arising out of the rigidities and dogmas of modernism. But is "arising," with its implications of empyrean goal and progress, quite the word? Is not a certain disorientation—induced by conflicting voices all declaring what is wrong without any consensual intuition of what is right—a hallmark of our era?

The book begins with the biographies of senior geographers, based on interviews conducted by the author. Here, then, is one more strain in humanism—attention to and respect for the individual. In the nature of the case, Anne herself cannot be featured among the biographical sketches—a big void in the literature, which I would like to fill but cannot here, except to say that she is a humanist in all the senses of humanism noted above, and therefore an ideal guide to the understanding of geography as a manifestation of the human spirit.

YI-FU TUAN

Acknowledgments

MY DEBTS ARE MANY, and my gratitude profound to literally hundreds of people who have been involved in producing this book and who have contributed to the International Dialogue Project 1978–88, some results from which are shared here. I especially acknowledge Torsten Hägerstrand, partner in the entire endeavor right down to the proofreading of the manuscript that became this book, and fellow members of the IGU Commission on the History of Geographical Thought, who have supported this idea from the start. I thank Suzanne Krueger, deft and dutiful secretary; Doris Nilsson, whose artistic contributions added so much throughout the decade; the studio personnel at AV-centralen in Lund; and especially Kurt-Åke Lindhe of Tetra-Pak, who graciously volunteered time and skill for video recording of interviews, often impromptu, in Lund and beyond.

Warm appreciation goes to friends and colleagues who have taken initiatives on recording interviews within their respective fields: among geographers, Ann-Cathrin Åquist (Sweden), Ove Biilmann (Denmark), Franco F. Ferrario (Italy and South Africa), Maria-Dolors Garcia Ramon (Spain), Robert Geipel (Germany), Arthur Getis (U.S.), Rudi Hartmann (U.S.), Ingvar Jonsson (Sweden), André Kilchenmann (Germany), Elisabeth Lichtenberger (Austria), Peter Nash (Canada), Joan Nogué i Font (Spain), Christina Nordin (Sweden), Philippe Pinchemel (France), Douglas Pocock (England), Joseph Powell (Australia), Allan Pred (U.S.), Shalom Reichmann (Israel), Joanne Sabourin (Canada), Terry Slater (England), Bruce Thom (Australia), Billy Lee Turner (U.S.), and Jeremy Whitehand (England); in other fields, Jonathan Bordo (Canada), Lucy Candib (U.S.), Grady Clay (U.S.), Inge Dahn (Sweden), Yngwe L. Ferdell (Sweden), Katrin Fridjonsdottir (Sweden), Pierre Guillet de Monthoux (Sweden), Karl Gustav Hjerpe (Sweden), John Humphrey (Singapore), Hans-Åke Jonsson (Sweden), Tore Nordenstam (Norway), Ligia Parra (Colombia), Madeleine Rohlin (Sweden), Inger Sondén-Hellquist (Sweden), Uno Svedin (Sweden), Carl Utterström (Sweden), and Erik Wirén (Sweden).

The recordings, as well as the career journeys of scholars from a wide variety of other fields, have added spark to courses and seminars at the universities of Lund, Cork, Clark, Paris I, Texas, and Ottawa during the 1980s. The central themes in this book have certainly been influenced, revised, and transformed through these discussions. Colleagues at each of these institutions have also offered critique and encouragement. At Clark, Bill Koelsch, Martyn Bowden, and Bob Kates offered valuable commentary; at Austin, Robin Doughty, Karl Butzer, and Terry Jordan afforded moments for reflection. At Lund, beyond Sölvegatan 10A, Birgitta Odén, Orvar Löfgren, René Kieffer, Pierre Guillet de Monthoux, and visiting scholar Edmunds Bunksé kindly offered critical suggestions. In fact, the story of geography in Sweden offers an excellent illustation of the main themes presented here.* In the final phases of getting the story into written form, the critical comments of colleagues and students at the University of Ottawa, Michel Phipps, John van Buren, Nancy Hudson-Rodd, Janet Halpin, Marc Vachon, and Cynthia Davey, were warmly appreciated. And at University College Dublin, Kevin Whelan's editorial help has been invaluable.

Financial support for the International Dialogue Project has come from diverse sources. In Sweden, the Council for Humanities and Social Science Research (HSFR), the Bank of Sweden Tercentennial Fund (RJ), and the Committee for Future-Oriented Studies (SALFO) funded salary and research endeavors throughout the period from 1977 to 1988. The Wallenberg Foundation provided funds for setting up an archive of project materials at the Lund University Library in 1986. At Clark University, the Tashahiki Fund and the Leir Foundation supported project activities in 1980–81, and the Higgins family provided funding for salary and office expenses from 1980 to 1985. The Rausing family and the Tetra-Pak Company offered personnel and expenses for all recordings conducted during the International Geographical Congress of 1984 (Geneva and Paris) and that of 1986 (Barcelona). The organizing committee of the 1988 IGC (Sydney) provided funds for four video-recorded interviews there, and the Faculty of Arts at the University of Ottawa supported the recording of three interviews in 1989. Donations of video recordings have come from various other sources (see the Appendix). To all, a heartfelt thanks.

My biggest debt is to my husband, Bertram Broberg, for reminders about classical texts, graphic illustrations, and shortcuts in PC word processing.

*This was originally designed as an integral part of this book but will now appear as a companion volume, whose working title is "By Northern Lights: On the Making of Geography in Sweden."

GEOGRAPHY
and the
Human Spirit

Introduction

                                        Numberless wonders
terrible wonders walk the world but none the match for man—
that great wonder crossing the heaving gray sea,
                        driven on by the blasts of winter
on through breakers crashing left and right,
  holds his steady course
and the oldest of the gods he wears away—
the Earth, the immortal, the inexhaustible—
as his plows go back and forth, year in, year out
with the breed of stallions turning up the furrows.

And the blithe, lighthearted race of birds he snares,
the tribes of savage beasts, the life that swarms the depths—
                                    with one fling of his nets
woven and coiled tight, he takes them all,
  man the skilled, the brilliant!
He conquers all, taming with his techniques
the prey that roams the cliffs and wild lairs,
training the stallion, clamping the yoke across
  his shaggy neck, and the tireless mountain bull.

And speech and thought, quick as the wind
and the mood and mind for law that rules the city—
  all these he has taught himself
and shelter from the arrows of the frost
when there's rough lodging under the cold clear sky
and the shafts of lashing rain—
      ready, resourceful man!
                        Never without resources
never an impasse as he marches on the future—
only Death, from Death alone he will find no rescue
but from desperate plagues he has plotted his escapes.

Man the master, ingenious past all measure
past all dreams, the skills within his grasp—
  he forges on, now to destruction
now again to greatness. When he weaves in
the laws of the land, and the justice of the gods

that binds his oaths together
he and his city rise high—
but the city casts out
that man who weds himself to inhumanity
thanks to reckless daring. Never share my hearth
never think my thoughts, whoever does such things.
—Chorus in Sophocles' *Antigone,* 5th c. B.C.
(trans. Robert Fagles, 332–71)

FOR ITS QUALITIES of ingenuity and skill in the conquest of nature, biosphere, space, and time; for its powers of speech and thought; and eventually for its responsibility in choosing good or evil, humanity was celebrated in Greek literature.[1] The *oecoumene,* or humanly cultivated world, posed an intellectual puzzle that invited speculation about the nature of things while also triggering exploratory voyages within and beyond Mediterranean shores. Like Chinese, Mesopotamian, and other early cosmographers, Greek scholars mapped out the known world into bands of decreasing familiarity from home to horizon, a spectrum fading from "civilization" to "barbarism."

As humans survey the earth from satellites in these latter years of the twentieth century, images of the *oecoumene* evoke worries about relationships between humanity and its terrestrial home. Waves of concern about global environmental changes surge across the horizon. Poignantly one recalls the *Antigone* picture of humankind, some focusing on its tragic import, many even rejoicing in its overthrow. Western civilization as a whole faces the humbling challenge of redefining its role within the anthroposphere, a challenge for which geography should ideally offer guidance.

The Occident has long cherished the belief that rational knowledge should provide the enduring beacon for life. Yet the world of contemporary science is fragmented, reflecting the fragmented worlds of its sponsors and audiences. Humanities and sciences are housed in separate, often mutually hostile domains. Whole realms of reality have been probed, explained, and set to work for the improvement of human welfare. Yet somehow the harvest from specialized research fails to deliver solutions to pressing problems. This book addresses this paradox in the noosphere as it explores the story of geography and humanism in the Western world.

The words *geography* and *humanism* have a chameleon-like character, revealing as much about their definers and their worlds as they do about any perennial truths. Interpretations of humanism reflect the time, place, and world-views of the interpreter. Despite the universalist aspirations of humanist lore, it cannot be fully appreciated until it is set within its own temporal, geographic, and cultural setting. Proclamations about the meaning of humanness—be it defined as *animal rationale* (rational animal), *homo sapiens* or *demens* (wise or insane), *zoon politikon* (social creature), *homo faber* or *ludens* (maker or pleasure seeker), each claiming generality transcending cultures, history, and environments—make little sense geographically until they are orchestrated with the more basic nature of dwelling. Humanists through the centuries have explored the nature of humanity, its passions and powers, while geographers have studied the earth on which humans, among many other life forms, fashion a home. For each facet of humanness—rationality or irrationality, faith, emotion, artistic genius, or political prowess—there is a geography. For each geographical interpretation of the earth, there are implicit assumptions about the meaning of humanness. Neither humanism nor geography can be regarded as an autonomous field of inquiry; rather, each points toward perspectives on life and thought shared by people in diverse situations. The common concern is terrestrial dwelling; *humanus* literally means "earth dweller."

What does it mean to *dwell?* Civilizations have varied greatly in their modes of understanding and dealing with the rest of the biosphere. In each civilization, the human spirit has sought to discern the meaning of earth reality in mythopoetic as well as in rational terms. The criteria of rationality and truth in every culture have always been derived from its foundational myths. Each civilization has a story to tell. The unfolding patterns of the earth around us invite a sharing of these stories as one essential step toward discovering mutually acceptable bases for rational discourse and wiser ways of dwelling.

The main ideas that I present here have emerged from a decade's effort to seek common denominators of concern that could serve as bases for dialogue and mutual understanding among practitioners of diverse specialties in science and the humanities. My central themes are journey and identity, storytelling and interpretation, individual experience and societal context, contingent events and broader historical movements. My chief concern is dwelling,[2] the most fundamental feature of humanness, and the role that scholars might play in improving self-understanding on the part of Western people and mutual understanding among the peoples of the earth. I seek to evoke awareness of contexts in which ideas and practices have unfolded and offer a set of interpretive themes which could elucidate connections between knowledge and life experience at different historical moments. Like all interpretations, these themes reveal as much about the author's own journey as they do about the people, places, and events being interpreted. They have unfolded and have been transformed and refined as the journey progressed.

It was to questions of integration in knowledge and particularly the challenge of building bridges between human and physical branches of geographic inquiry that Torsten Hägerstrand and I addressed our efforts during the years 1978–88. Geography was an ideal focus of attention, as a discipline whose domains of inquiry overlap the natural sciences, the humanities, and the social sciences, yet one that has persistently claimed an integrated perspective on reality. The institutional separation of physical and human geography which occurred in Sweden and elsewhere at midcentury had enabled far more incisive research into specialized subjects, sharper analytical methods, and closer association with colleagues in neighboring disciplines. *Ceteris paribus* (other things being equal), facets of geosphere and biosphere were scientifically explored with colleagues in natural science. Human geographers, aligning their methods with positivist procedures then in vogue among social scientists, analyzed facets of the anthroposphere, also in *ceteris paribus* terms, often deriving normative guidelines from spatial analysis for regional and urban planning. After a few generations of research conducted in specialized subfields, many geographers recognized the need for more integrated approaches. Now that global issues again claim attention, they realize that despite spectacular analytical feats of systems-analytical inquiry within both "human" and "physical" branches of the discipline, the *oecoumene* still presents unresolved dilemmas.

The term *integration* initially struck me as being simultaneously ambiguous and daunting. The volume of prose dedicated to this issue, from the philosophy of science to the sociology of knowledge and treatises on language and power, and on history and politics, seemed overwhelming. In the literature surveyed, the word *integration* itself evoked ambivalent emotional responses veering from eulogy to panic. Despite the positive connotations the term may hold in art, architecture, mathematics, and engineering, it seemed to imply in this particular context a managerial, or "top-down," solution to the issue of research specialization. Functional specialization was probably an inevitable trend in science, one with potentially invaluable merits. Ideally, I argued, if each specialist could explore connections between knowledge and experience within his or her own field of expertise and could engage in dialogue with other specialists on precisely these connections, then eventually some bases for mutual understanding and better integration of research results on substantive questions could be possible. With characteristic largess, the Swedish Council for Humanities and Social Sciences approved the idea of an international dialogue project, some of the fruits of which are shared in this volume.

The core of our experiment was a series of video-recorded interviews with senior and retired professionals in various fields where stories were told of career experiences, the dream and reality of major projects, and the circumstances in which ideas were inspired, developed, and tested.[3] These videotapes were then shown to small groups of specialists in different fields as catalysts for discussion and reflection among people who otherwise had little occasion for dialogue.[4] Between 1978 and 1988, 150 individuals from a wide variety of fields generously shared their career stories in either written or video-recorded form; it was from studying these texts that the main interpretive themes of this book emerged.

The Practice of Geography (1983) offered a selection from these career stories, each of which was unique but gave suggestions about potential common denominators of interest concerning creativity and context, thought and practice. A trilogy of themes, *meaning, metaphor, milieu,* was suggested as a potential framework within which common denominators of interest among specialists in the practice of geography could be discerned. *Meaning* refers to vocational skills, talents, and work preferences; *metaphor* points toward cognitive style, or basic world-view, underlying research models and paradigms; and *milieu* embraces not only those environmental circumstances deemed relevant in the formation of a person's orientations to meaning and metaphor but also those enduring constellations of public interest that geography has sought to elucidate, through disciplinary thought and practice (see Chapter 1). These three intertwining themes enable one to appreciate the uniqueness of each individual's career journey and simultaneously to discover general processes involved in the relationships between scholarly practice and its societal context.

As many a researcher knows, once a comprehensive way of seeing the world begins to crystallize, it then becomes the mirror in which one seeks to elucidate more and more phenomena. For me, the trilogy of meaning, metaphor, milieu not only served as a satisfactory framework for teaching courses and seminars on the history of geographic thought but also provided a fresh perspective on empirical and practical questions.[5] The attraction of this framework lay in its integral character. I became attached to it, eager to test it in various settings, impatient about details that refused to fit. But the process itself shed light on the question of how knowledge is integrated. At each step along the way, there had been tensions between two opposing desires: on the one hand, recognizing and affirming the integrity and possible uniqueness of individual careers and events; on the other, reaching toward common horizons and general patterns. The moments of sheer delight were those when glimpses of common horizons and shared experiences led to a deeper, better integrated understanding. Others might have made even more extensive journeys toward their own thematic trilogies or favorite mode of grasping things integrally. How might we negotiate our "integrated" versions?

Further scrutiny of career accounts and continued dialogue resulted in some reconsideration of the original interpretive framework. As a conceptual scheme, there was little doubt about its heuristic value, even if confusions and ambiguities often arose in its application to particular cases. It afforded a cross-sectional (synchronic) interpretation of texts in context. But what of the diachronic? Surely one of the main values of studying life histories was the opportunity to catch a glimpse of the flow of events, changes of course, and continuities and discontinuities, rather than just interrelationships between thought and context at particular moments. Most illuminating for authors was the discovery of resonances between their journeys and those of others, of similarities and differences in their attitudes toward external events that occurred during their working careers.[6] The social and academic milieus in which these authors had pursued their careers had witnessed many radical changes from the optimism of pre–World War I days through the Depression years to another world war, and then to the great excitement over postwar reconstruction and regional development. How might one characterize the changing intellectual milieus surrounding those career journeys? Could these diverse individual journeys be related to the diachronic flow of external events and trends?

At moments of intellectual impasse, few things are more welcome than a compelling diversion of attention. For me it came through an invitation from the late Walter Freeman, secretary of the International Geographical Union Commission on the History of Geographic Thought, to write a bio-bibliographical study of Edgar Kant, and through another invitation from Spanish colleagues to speak on the historical foundations of humanism in geography at a workshop in Madrid in 1985. Just as previous diversions had helped clarify and refine this meaning-metaphor-milieu framework, so these two opened up the question of diachronic flow in career histories and in the development of ideas and practices over time. A fresh trilogy of themes emerged, *Phoenix, Faust, Narcissus,* a mythopoeic rather than rational mode of interpreting knowledge and life experience, and a wider synthetic frame within which all previous stories could be elucidated. In historical perspective, the essential message of humanism could be regarded as an emancipatory cry, in dialectical (and potentially creative) tension with the rational quest for integrating knowledge. Within the overall drama of dwelling on earth, it might be best defined as the *cri de coeur* of humanity, voiced wherever its integrity was threatened or its horizons dimmed (see Chapter 2).

Throughout the historical record there were moments of breakthrough, partial integrations of knowledge and understanding, that shone forth in the ashes of former certainties. New discoveries in various realms of knowledge and life were construable as *Phoenix* moments—the birth of a nation;

the stirrings of revolution; new levels of awareness in art, music, literature, or scientific explanation. Once structured and institutionalized—integrated as model paradigm, for example—*Faust* entered the drama—ideas and practices became normalized, justified, sometimes defended as the orthodox way of doing things. And eventually, as in our own day, the reflective *Narcissus* could generate two kinds of response: the love-hate relationship to one's Faustian folkways, or the openness to allow new alternatives to emerge.

Suddenly a whole new perspective opened up, not only on the research challenge of integrating knowledge, but also on life histories that I had been reading through the favored filter of my interpretive framework: meaning, metaphor, milieu. The energy invested in analyzing components evident at various stages of an author's career was illuminating, but all the while it was clear that the most fascinating story was the unfolding of their life journeys. Back and forth through informal and formal encounters during this decade in Sweden, from the broad sweep of Western history to the individual life story, the fundamental drama of Phoenix, Faust, Narcissus continued to resonate.

If only I could have glimpsed such an interpretation at the beginning, perhaps the last decade's journey would have been smoother. In retrospect, there was something of a Phoenix mood at the outset: the prospects for potential dialogue between humanist and social engineering (or benign technocrat) perspectives on geography and life experience seemed indeed an exciting challenge. It was easy to initiate dialogue where communication was face-to-face, and where all kinds of questions could be aired spontaneously. It was quite another task to justify dialogue as part of a reportable research project. Faustian protocol required a caginess about latent "no trespassing" signs between disciplinary territories. The tensions between process and product in those early years seemed insurmountable, but it could now be elucidated, mythopoeically, as tensions between Phoenix and Faust. The entire dialogue effort could be reconstrued as Narcissus visiting the waters of Helicon.

Narcissus has returned from the pilgrimage refreshed, having now caught glimpses of the drama in all that has transpired, understanding—even if not affirming—the vicissitudes of academic history and context, and ready to welcome whatever even newer Phoenix the future has to offer.

Part 1 of this book introduces two sets of interpretive themes: meaning, metaphor, milieu (Chapter 1) and Phoenix, Faust, Narcissus (Chapter 2). Part 2 explores the larger drama of geographical thought within the general social and intellectual contexts in which disciplinary ideas and practices have unfolded in Western history. Four root metaphors are presented as *dramatis personae* in Western intellectual history, and the following chapters trace the story of how each has negotiated the interests of meaning and milieus over time: world as *mosaic of forms* (Chapter 3), as *mechanical system* (Chapter 4), as *organic whole* (Chapter 5), and as *arena of spontaneous events* (Chapter 6).[7] The concluding chapter returns to the questions of specialization in research and the problem of integrating knowledge, relating these to the tensions of local and global horizons. It points toward the need to reach beyond the realm of Western myth and metaphor and understand ways in which other cultures have construed the nature and meaning of their biophysical environments.[8] It also draws some conclusions regarding geography's potential role in clarifying the contemporary challenge of terrestrial dwelling. The Appendix contains a full listing of International Dialogue Project recordings (1978–88), references to which will be given by serial number in the body of the text.

This story of geography and the human spirit should resonate to the experiences of my own generation, who were students in the 1960s and have been witnesses to events of the past few decades. Dialogue was a winged word during those heady Phoenix days; new life was in the air, and there was an eagerness to question virtually everything and to experiment with new ways. This emancipatory élan soon became institutionalized and normalized, and Faustian fences were drawn around discrete territories. During the 1980s, many indeed became cynical and disillusioned. Narcissus is languishing. Meanwhile, new challenges surface on the horizon. It seems indeed to be high time for emancipation from the encrustations of institutionally defined structures—those Faustian frames created by our own generation with zeal and conviction—which are no longer appropriate for the challenge of understanding humanity and earth. This is surely one lesson that geographers have had to learn throughout the centuries. The human spirit in the Western world beckons us to transcend those structures set in place to guarantee "freedom from" oppression and to direct our energies toward "freedom to" dwell. Cypran Norwid's poem that stands as epigraph for Jerzy Andrzejewski's twentieth-century novel *Ashes and Diamonds*, echoes the Phoenix hope:

> Shall it just mean ashes and dark turmoil
> Dashing about in a stormy chasm
> Or among the embers does a diamond shine
> Herald of daybreak and invincible light?

Part I

Phoenix, Faust, Narcissus

Chapter One

Meaning, Metaphor, Milieu

Life Journeys in the Practice of Geography

> It is strange that some outside pressure was needed for me to see that the obvious point to begin when trying to link the outer and inner worlds is where you can look in both directions: with oneself.
>
> —Hägerstrand (in PG, 256)

AUTOBIOGRAPHY is surely a powerful catalyst for dialogue. Each story is ultimately unique: each unfolds through different circumstances, posing its own range of curiosity about the interplay between creativity and context (Koestler 1978; Chargaff 1978). Readers may begin to think critically about their own life journeys, and intellectual historians may be prodded to question conventional certainties about the interplay of events and contexts. For the narrator, too, autobiography offers an occasion for discovery. Interpretations of these stories inevitably reflect the ongoing curiosities of the reader.[1] This chapter describes one way of interpreting career journeys in geography, given the goals outlined in the Introduction.

The central challenge for the International Dialogue Project was to shed light on issues of knowledge integration, with particular emphasis on relationships between human and physical branches of geography. Functional specialization among research fields was patently associated with a general fragmentation of thought and life, in academe as well as in society. Instead of confronting this complex situation (and all the managerial and philosophical issues involved) head-on, it seemed safer to try first to understand scholarly ways of life and the interplay between research practices and their societal contexts. Specialization, after all, seems to be a fact of life and a tendency most likely to intensify; gains on research frontiers apparently outweigh those of relevance to either ongoing societal problems or issues of integrating knowledge. The specialist is surely not just a bona fide member of an intellectual caste; each belongs to a disciplinary way of life, a professional vocation that involves emotional, moral, aesthetic, and cultural preferences as well as choosing priorities among ways of thought (Geertz 1983; Bourdieu 1977). Nor can the academic specialist's thought and practice be understood until one also understands vicissitudes of ideology, place, and time among sponsors and audiences for a particular kind of expertise. It was in hopes of discerning some experientially grounded bases for mutual understanding and some common ground for critical evaluation of disciplinary ideas and practices that the narrative mode was selected. The dialogue process began with this first step in the direction of bridge building, not only between academy and society, but also among different cultural, linguistic, and national traditions.

And why not begin at home, as it were, within geography itself? The geographer's career journey especially highlights interactions between human intentions and hopes and environmental contexts through which careers unfold (Löffler 1911; Schouw 1925; Taylor 1958). A motley crowd we are, ranging in research curiosities from glacial morphology to environmental perception, many finding it easier to communicate with colleagues in other disciplines than with one another. Senior and retired colleagues who still advocate more integrated modes of disciplinary practice conducted their careers through variegated contexts: war and peace, depression and boom, migrations and revisions of territorial boundaries, each country offering its own set of challenges. Their expertise as teachers, surveyors, explorers, and researchers was welcomed in particular nations, and they have sought to fulfill these tasks with energy, skill, and devotion. Then, as all the details on each nation's world map got progressively filled in, and as matters of domestic housekeeping, industrial location, and administrative regionalization became resolved, the geographer's traditional task seemed to become dispensable. Either the job was done or else it could be assigned to experts in other fields, such as law, demography, economics, or the civil service. So geography, as an integrated field of scholarship in the late twentieth century, having witnessed its offspring leaving home and taking on new names, was facing an identity crisis and fragmentation of its diverse branches.

Although the dialogue project extended to many fields and themes, attention was primarily addressed to the texts of senior geographers. A hermeneutic approach, rooted in historical science and critical theory, and articulated especially by Wilhelm Dilthey and Jürgen Habermas (Dilthey 1913–67; Habermas 1968, 1976), seemed most appropriate.

> An infinite wealth of life unfolds in the individual existence of particular persons in virtue of their relations to their milieu, to other persons, and to things. But every single individual is at the same time a point of intersection for structures that permeate individuals, exist through them, but extend beyond their lives. These structures possess independent existence and an autonomous development through the content, value, and purpose that is realized in them. (Dilthey [1913] 1967, 7: 134)

Dilthey proposed the idea of a "community of life interests" as a precondition for objectivity among practitioners of the *Geisteswissenschaften* (cultural sciences), emphasizing two major challenges: (a) discovery of a *language* to permit intersubjective communication and mutual understanding and (b) acknowledgment of an individual life story as prototype for the diachronic connectivity of knowledge and experience (Habermas 1968, 140–86). What better data source, then, than career histories for insight into both? The process of dialogue over life journeys had intrinsic pedagogical value and afforded ample opportunity for exploring common ground between those who favored "scientific" and those who preferred "humanistic" modes of interpretation. But evidence abounded that "hermeneutic understanding must inevitably employ general categories to grasp an inalienably individual meaning" (Habermas 1968, 159).[2]

It was from these career stories and from wider readings in the philosophy of science that the main interpretive themes of this chapter were discerned: *meaning, metaphor, milieu,* and *horizon*.[3] *Meaning* points to vocational preference and work skills; *metaphor,* to cognitive style. *Milieu* evokes not only circumstances deemed relevant in the formation of a scholar's choice of field or style of practice but also the human (societal) interests to which the discipline as a whole has addressed itself. Janus-like, these themes point in two directions. They serve as searchlights that play on the practice of geography, highlighting the perennial tensions between scholarly practice and its environment, between the personal preferences, talents, and goals of individuals, on the one hand, and external challenges and circumstances, on the other.[4] Distinct as they may appear to the analytical gaze, in practice they interweave within the *horizons* of particular periods and settings (see PG, 62–66, 186–195; also NL). These themes also interweave and change through a scholar's career journey; they serve to identify conflicts between individual and societal goals. They point toward common denominators of concern among practitioners of diverse research fields and together form a framework within which to evaluate the practice of geography in the changing contexts of the late twentieth-century world.

Meaning

Why geography? A twelve-year-old Swedish youth returning home from holidays in the south of England in 1914 "on a blacked-out ship overloaded with returning Russians" asked himself, "How could people be so stupid as to start a world war?" He concluded that "they simply do not understand each other's history and geography," and thereupon chose a career in geography (William-Olsson in PG, 155). Josep Iglésies discovered his geographical interests while hiking in his homeland territory: "I wasn't simply a trail-blazer. . . . I asked myself many questions and built up a firm interest and love of Catalonia, not in abstract terms, but with the desire to touch it and get to know it down to the last detail" (Iglésies, GGS2, 5). For William Mead it was the Vale of Aylesbury which first offered inspiration:

> I can recall moments (recent as well as long ago) when time was suspended in it. It is a green vision, luminous in early June haze, best perceived floating downhill into the warmth of the Vale on a bicycle or riding along a bridle path with near-ripe grasses tapping the toes in the stirrup. It is just before the hay is mown and just before Matthew Arnold's "high midsummer pomps" take over. Then, the sense of place is at its most personal. (Mead in PG, 47)

The aesthetic appeal of landscape exploration and the emotional sense of belonging to place have ranked high among the sirens that have called people to the practice of geography. It was a field that offered them an opportunity to link home and horizon, microcosm and macrocosm. Long travels and musings on the world and its cultures no doubt sowed the seeds of geographical curiosity for Clarence Glacken, but it was a casual invitation from Carl Sauer that brought him back to a formal position in the discipline at Berkeley: "I knew definitely I would [join the department] when he told me what office I could have" (Glacken in PG, 28). "I must admit," Aadel Brun-Tschudi confided, "that most of what I have done as a geographer has been the result of happenstance, from invitations received from others," yet she claimed convincingly, "to me geography is a vocation" (Brun-Tschudi, G1, 5). "I should have preferred to be a banker, or a medical doctor," Wolfgang Hartke volunteered, "but there was something else. . . . I wanted to be mayor of a big city. The reason? I was a very small person and I wanted to have many contacts with all the people of the city" (Hartke, G6, 4). "It was by mere chance that I became a geographer. But even if I had become a medical doctor I would have had the same scientific interests as I do as a geographer. What influenced me? Two unhappy love affairs, a fight with my father, leaving the family, those were the reasons why I became a social geographer and not a classical geographer" (Hartke, G1, 4).

"From the very beginning," G. J. van den Berg stated, "I perceived that geography could be really instrumental only as human or social geography" (Van den Berg in PG, 210). "In geography there are always many black boxes," Hartke remarked, "but the more one tries to make it practicably useful, the better it can serve humanity" (Hartke in PG, 235). "I'd rather put my energies where things can help," said Robert W. Kates, "although it is perfectly credible and marvellous that people work on esoteric subjects"

(Kates, G23). For many geographers, such concern about social relevance may be uninteresting, and their central vision may be to bring the discipline into the ranks of proper science. "I believe in calling a spade a spade," said Richard Chorley; "I mean if we're going to use—manipulate—geography to bring about the sort of world we want to bring about, OK, let's say so. But let's not kid ourselves that we are trying to look objectively at the operation of processes and the sorts of forms which result from them" (Chorley, G25). For many, however, it was neither the status of geography as a science nor its actual or potential relevance to societal problem solving that made the field attractive as vocation. "I always looked on teaching as my primary responsibility," John Leighly wrote; "original work and writing for publication were things to be done in free time" (Leighly, G14). And for a few, what was most significant was that geography opened up a vast challenge of inquiry and reflection on the connections between humanity and nature: "The personal value of producing a scholarly work is the reward of any form of art, fulfillment of the challenge of fashioning a new creation in the midst of what often seems a world of chaos, without form and void" (Koelsch 1970, 32; see also Spate 1960).

When people reflect on their career journeys, implicitly or explicitly seeking threads of continuity and meaning in their experiences, their thoughts often turn to the actual works in which their life energies were invested. Scrutiny of oral and written statements leads to the conviction that work itself—call it "vocational meaning," if you are existentially inclined; "social role," if you are a functionalist; "labor," if you are a materialist—is a compelling common denominator among people from all walks of life. Academics are no exceptions: teachers tend to understand teachers; researchers appreciate scientific rigor and discovery; social activists and planners get on each other's wavelengths (or nerves) with little effort; the scholarly writer will appreciate the art of literary creation wherever it appears. One central theme in the interpretation of geographic thought and practice is *meaning*. Four distinct constellations of vocational meaning are discernible in the career accounts of senior colleagues (fig. 1):

Poesis (πσ)[5] denotes the evoking of geographic awareness, critical reflection, discovery, and creativity. It elicits curiosity and insight about relationships between humanity and the physical earth in themes such as culture and landscape, sense of place, nature symbolism, or the history of ideas. It may take the form of playing with ideas, simulation exercises, speculations about language and power, or the ethics of applied geography. The poetic dimension should ideally address the critical and emancipatory interests of all other practices.

Paideia (literally, "education," "formation") refers to teaching, lectur-

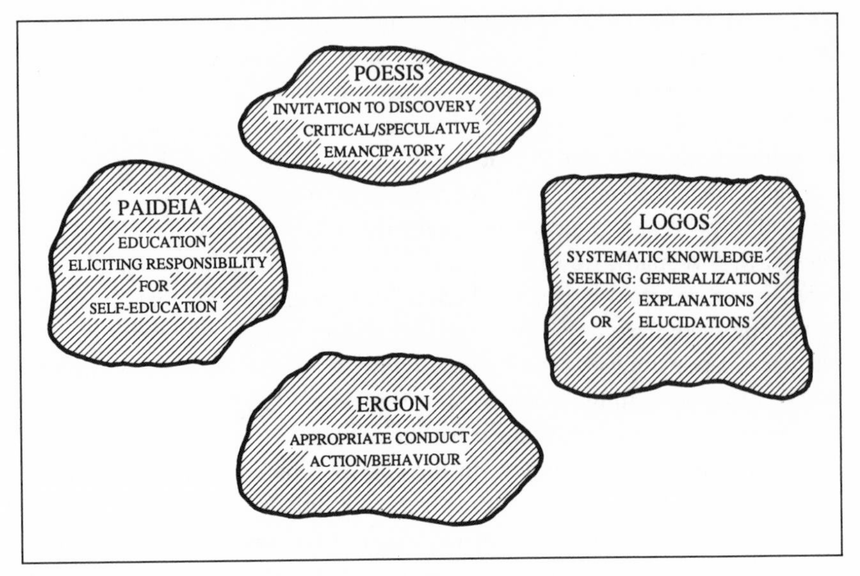

Fig. 1. Constellations of vocational meaning

ing, designing and guiding field excursions, and counseling students. It may be expressed in textbooks or teachers' manuals, in the organization of workshops, and strategies for curricular reform. Often, it seems, those who espouse *paideia* as central vocational meaning also assume responsibility for archival, editorial, and other projects in the circulation of information.

Logos (literally, "systematic organization of thought") is expressed via the promotion of analytical rigor, objectivity, and science making—be it in relation to soil samples, transport networks, or environmental perception. Scientific *logos* seeks explanation rather than description; generalizations about spatial systems and the functional organization of space, for example, are deemed far superior to the understanding of particular places or regions. One's goal is to make geography a scientific field. This does not mean that humanists eschew generalizations, but rather that their goals are somewhat different. Humanist *logos* seeks elucidation and understanding rather than causal explanation.

Ergon (literally, "work," "action") denotes efforts to render geography relevant to the elucidation and solution of social and environmental problems. Issues assume central importance. Some applied geographers focus on spatial and environmental aspects of problems, and they are generally less concerned about disciplinary orthodoxy than about problem solving. Practitioners of applied geography often adopt a social reference world that only partially overlaps that of their academic colleagues. Some assume administrative responsibilities in universities, research councils, or civil life, hearkening to the call of *ergon*.

There are many other elements of vocational meaning which individuals acknowledge as important in their own practice of geography—for example, carrying on a family tradition; the aesthetic delights of mountain climbing, hiking, and travel; musing on maps; becoming engaged in issues of war, peace, the struggle for national independence, or defending the national or imperial flag. Such motivations cut across and do not fit neatly into the four rubrics outlined above. In the interests of international and intercultural comparisons, however, I suggest that these four clusters of vocational concern, *poesis, paideia, logos,* and *ergon,* might constitute a set of general themes applicable to diverse languages and specialties. They may be considered as social practices that reflect external circumstances, such as public attitudes and research sponsorship, as well as individual competence or talent. They could thus provide common denominators for intra- and interdisciplinary discussion about priorities in academic practices at any particular time.

Rare, of course, are the individuals who could be identified exclusively within one of these rubrics. Most have combined several at different moments in their careers. Values change over the course of a lifetime, and most geographers have been open not only to changing options but also to combining different activities in creative ways. There was a student who entered Lund University in 1937 fully convinced that the last thing in the world he would choose would be the teaching profession. Forty years later, when asked about geography's role in future society, he laid emphasis on the educational one (Hägerstrand, G9). Teaching has been a duty, if not a choice, for most geographers; tedious as that may have been at times, many acknowledge how valuable it has been as complement to research and other activities. "Undergraduate students are fresh," wrote Glacken, "capable of asking most difficult questions, often jaunty, lazy at times, which can be an advantage because a professor's words and writings are often not *that* memorable" (Glacken in PG, 33). "For inspiration, support, and challenge, I go to my students," wrote van den Berg at the seminar on creativity and context in 1978, "especially when they are on field work. . . . The students represent a significant context for fostering and maintaining my personal growth." "Young geographers should think of their discipline as an integration of research and teaching, of university life and participation in public life," Jacqueline Beaujeu-Garnier advised (Beaujeu-Garnier in PG, 151). *Ergon, logos, paideia,* and *poesis* offer conveniently distinct categories for interpreting texts, but they are certainly not separable or mutually exclusive in the lived experience of scholars.

They do, however, evoke another important aspect of vocational practice, namely, the mode of discourse. Someone who performs brilliantly in the lecture hall or on a field excursion with students may have neither the time, energy, nor motivation to communicate in writing ideas that had such oral appeal, face-to-face, in particular settings. Before midcentury, many geographers practiced the arts of *poesis* and *paideia* orally. As far as *logos* is concerned, some individuals think visually, finding delight in graphic or mathematical modes of symbolizing complex problems or seeking solutions. Some reason in syllogisms, others with maps or statistics, others in the tropes of literary style. For *ergon,* the preferred mode may be action and experiment, political persuasion or participation. The practice of geography has witnessed many changes in the values ascribed to oral versus written modes of expression, to cartographic or mathematical modes of description, to literary or graphic styles of reasoning, to ivory tower or marketplace—changes fully as dramatic, in retrospect, as shifting paradigms, theories, or methods. The mode of discourse itself mirrors both internal preferences of practitioners and external propensities of sponsors and clients.

For one certainly cannot assert that the allocation of time and energy to different tasks has been a matter of personal choice; in most cases it has been a matter of externally defined opportunity or administrative design. To hold a position as a geographer in a university department or government office implied duties prescribed within the prevailing division of labor. The generation retiring in the 1980s had to combine teaching and research, and several had additional external consultancies. It has often been extremely difficult to find the time and place for reading and reflection. In the early years of discipline making, a geographer indeed had to become a jack of all trades. Walter Freeman confesses frankly that during the late 1940s at Trinity College, Dublin, "I *was* geography!" (Freeman, G7). And in the later phases of virtually all career stories told, the most salient influences were those related to research grants and the availability of funds.

This interpretive theme, *meaning,* affords an avenue of inquiry into the interplay between individual vocation, talent, or preference and the externally defined job definitions within which people conduct their activities. From an external viewpoint, one could observe the changing status and national investment in these different works during the disciplinary period. Each school no doubt has its own record of fluctuations over time in the status and demand for humanists, teachers, scientists, and planners. Ministries of Education may regard these as budget items competing for limited resources; from the practitioner's viewpoint, however, they may be seen as complementary and mutually enriching foci of vocational meaning.

An examination of doctoral dissertations in geography written by individuals who achieved docent status[6] within Swedish universities during the twentieth century revealed the following pattern (fig. 2).[7] In Sweden, as indeed in most major schools in other countries, the increasing demand for scientific research or practical applications of geography since World War II has certainly diminished the perceived status of teaching and critical reflection.[8]

These categories of meaning thus point in two directions: to the social structuration as well as the existential value of geographic practice. They indicate the personal experiences of geographers as knowing subjects and the shared designations of role and activity which have prevailed in various contexts. One horizon opens onto a vista of exploration into the genesis of geographic awareness, vocational choice, competence, and curiosity; the other faces onto that fascinating world of institutional arrangements surrounding geography's place in the university and in society. Beyond the worlds of professional task definition, however, there are intellectual common denominators shared by people in different walks of life: root metaphors of reality which offer other potential bases for mutual understanding among geographers.

Cognitive Style and Metaphor

> Like every geographer I love maps. But the great adventure of my scholarly life has been to try to transcend the map. I see, almost literally, the opulence of the world as a *moiré* of processes. (Hägerstrand in PG, 239)

Searching for the sources of his own concepts, Torsten Hägerstrand described his career as "a lifelong marriage . . . which has exposed me to an enduring tension between attraction and repulsion from the moment I was engulfed in the academic version of the discipline" (Hägerstrand in PG, 238). Peter Gould called *Geographia* "a delicious and seductive wench" with

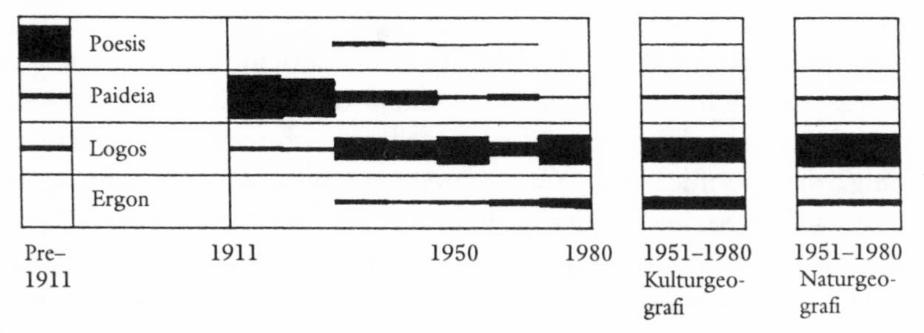

Fig. 2. Changing emphases in the practice of geography. (Illustration from a study of Swedish doctoral dissertations, 1880–1980.)

whom he confessed to having had a passionate love affair since the age of fifteen (P. Gould 1985, xv). And Karl Gustav Izikowitz, a social anthropologist, wrote of his life as a "pruned tree," whose trunk grew from a particular seedbed and whose branches and profile took shape through various winds and weather. In later years he pondered why the crown was so unevenly shaped. "Only a wildgrown tree standing in a protected spot can manage without a crown, but this is rare. Ambitions are a kind of fertilizer: if you give too much the plant becomes sick" (Izikowitz in GN, 108). Anthony Padovano, a theologian, summarized his life journey in geographic terms: "It begins with the scene of a young boy standing alone by railroad tracks in rural Kentucky. Every image that follows is one of movement: a journey to Birmingham and a road to Rome, an automobile ride into frenetic New York, a return along back roads to Darlington, the rush of rivers at Pittsburgh, an aircraft that circles the city of Seattle and the waterways that lead into the wilderness" (Padovano in Baum 1975, 233).

Diverse and intriguing are the similes and metaphors used in personal biography (Schouw 1925; Christaller 1968; Taylor 1958). Psychologists and intellectual historians ponder variations among individuals of different disciplinary backgrounds, of different periods and milieus. The correspondence between an author's way of telling his or her own life story and the preferred cognitive style adopted in published works is remarkable. In some cases, the correspondence seemed striking. "A career," wrote William Warntz, "like any progression, is comprehensible only within a frame of reference—a coordinate system—against which its directions and velocities are measureable. Moreover, the frame of reference, if properly chosen and described, may prove to be of more interest and importance than any particular paths traced within it." Here was a person who rejoiced that "the scope [of disciplinary research] was broadened to include the investigation of the various kinds of social energies and their interplays." His aim was to establish a unified social science and to show that it and physical science were "but mutually related isomorphic examples of one generalized logic" (Warntz in Billinge et al., 1984, 135).

Some see their careers as products of circumstance, each phase construed as flowing from the chance interactions of external and internal influences. Others see their life journeys as rivers with tributaries, eventually joining some cosmic ocean. One retired geographer drew a straight line with an arrow pointed vertically, with punctuations marking moments of career change, marriage, projects begun, or the outbreak of wars. Another depicted a promenade in the woods, with cartoonlike inserts describing clusters of research interests and unexpected catalysts for change in his own practice of geography. Yet another drew a river basin, with converging valleys bearing the labels of thought currents, and asterisks positioning key individuals as well as books and articles that had exercised significant influence on his orientation. Some communicate in pictures, some in words; often, the cognitive style apparent in an author's published works is mirrored in the one chosen for telling one's life story (Nash 1986).

It is indeed conceivable that each scholar develops a distinct metaphorical style in the course of a career (Polanyi 1958; Gardner 1983). A vital distinction needs to be made, however, between metaphors facilitating the process of discovery and those used in communicating results. Sound, touch, smell, and vision all play a part in the molding of geographical awareness, and metaphors of discovery are intimately linked with a scholar's childhood experience. How a scholar writes in scholarly journals may not reflect how he or she has come to understand the subject; rather, it may reflect his or her perceptions of what an audience would grasp. To understand the published works of any author, one needs to understand those cognitive styles that prevailed within the discipline and were favored by editors during successive periods.

Conventional modes of describing geographic thought and practice characterize the field in terms of methods, epistemology, or substantive foci of research interest (J.N.L. Baker 1963; Dickinson 1969). Contextual approaches point to institutional and ideological features of disciplinary practice (Stoddart 1981). Each of these approaches sheds particular light on disciplinary thought, but virtually all yield evidence of a growing functional specialization and eventually of communication impasses among various fields of inquiry, namely, a restatement of the problem with which this entire project began. Some alternative approach is needed if one is to discern potential common denominators of cognitive style.

Metaphor, as I argue in Part 2, offers a uniquely valuable route toward elucidating the intellectual *Zeitgeisten* of successive periods, a route that unmasks more than does a scrutiny of paradigms, theories, or models. Of the enormous range of potential metaphors of reality, the four "root metaphors" presented in S. C. Pepper's *World Hypotheses*—formism, mechanism, organicism, and contextualism—provide one approach to the discernment of distinct cognitive styles in the practice of geography (Pepper 1942).[9] Each of these four world-views projects a distinct interpretation of reality; each could be regarded as parent to whole families of theory, paradigm, and model of research inquiry. In bold strokes this approach illustrates the case for pluralism. Each of the four root metaphors, Pepper claimed, has its own "theory of truth" and categories of inquiry: hence the products of research conducted within the framework of one could not be subjected to criticism via the categories of another. This schema therefore provides a forum where the relative strengths and limitations of different theories and methods could be articulated, and where the credibility of those periodically launched imperialistic manifestoes for the discipline as a whole could be assessed (fig. 3).

Mosaic, or *map,* seeks to represent, as accurately as possible, realities presented to the senses, resting its cognitive claims on a correspondence theory of truth (see also Chapter 3). The map projects a dispersed world-view: reality is construed as a mosaic of patterns and forms. Beneath each pattern, *formative* processes are at work, for example, cycles of erosion, ebb and flow of the tides, demographic transitions, or modes of production. Mosaic stands as central root metaphor for geography's chorological tradition, the most practiced of all four metaphors. It emphasizes analysis more than synthesis, appeals to the status quo or the democratically inclined; it provides bases for mutually beneficial exchange between geography and a host of other disciplines concerned with form, classification, and categorization, whether in psychology, biology, sociology, art history, or brain research (Judson 1980).[10]

Mechanism shares with *mosaic* a keen concern about analytical rigor and seeks precise and measurable data on which to ground its world picture (see also Chapter 4). Its image of reality as a whole is an integrated one: causally interrelated systems and interactions among systems constitute the prime curiosity. The cognitive claims of this root metaphor rest on a causal adjustment theory of truth. Mechanism stands as root metaphor for geo-

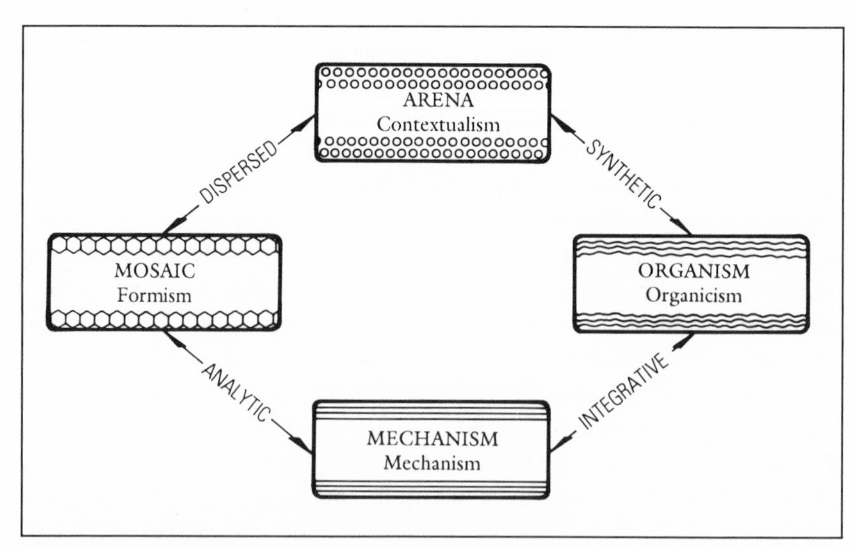

Fig. 3. Four root metaphors

graphy's "spatial-systems" approach, best illustrated in the push-pull of migration flows, positive- and negative-feedback processes in urban or hydrological systems, and gravity models of commodity movement.[11] Ideologically speaking, it has been observed that social reformers and revolutionaries have often used mechanistic models to show what they perceive as (a) injustices or inefficiencies in the status quo and (b) their vision of society in the future (H. White 1973).

Organism,[12] root metaphor of an organicist world-view, rests its cognitive claims on a coherence theory of truth (Chapter 5). It seeks synthetic accounts of reality, emphasizing unity in diversity, dialectical tensions yielding ultimate resolution within an organic whole. Organismic metaphors for reality as a whole may be found in most human civilizations and the Western (post-Hegel) applications of the idea must be regarded as only one special case, reflecting cultural and political circumstances of Western history.[13] Ideologically speaking, it has appealed to monarchists and anarchists, revolutionaries and conservatives, materialists and idealists. Aesthetically, too, the metaphor has appealed to humanist, poet, and painter, as well as to scientist and planner.

Arena (lit. "sand") is proposed here as the root metaphor of contextualist orientations in geography (Chapter 6). Its world image is that of a stage on which spontaneous and possibly unique events may occur. Arena attempts to provide a synthetic account of such unique cases, elucidating each in reference to its own particular context. Its cognitive claims are based on an operational theory of truth which, as in American pragmatism, allows for each case to be evaluated and judged on its own terms.[14] In North America, at least, one might credit contextualism for the motivation underlying that transition from "spectator" to "participant" styles of inquiry for a number of geographers (Rowles 1978; Kearns 1988; Folch-Serra 1989). Insofar as it helped unmask the culturally relative nature of thought styles and modes of discourse, it could also be regarded as herald for more hermeneutic sensitivity among researchers.

The extent to which these four world-views elucidate differences of cognitive style among the authors whose works we scrutinized varies from one individual to another. The schema works better at the macro (external or societal) level, allowing one to grasp some connections between intellectual currents and concurrent ideological and material circumstances, than it does at the level of particular individuals. Nor indeed was such stereotyping of individual authors intended in Pepper's work. Most creative scholars, he pointed out, avail themselves of more than one root metaphor in the course of a career. Only the rare dogmatist clings to one throughout. However, one could assume that people who showed consistent preferences for the same root metaphor would more easily understand one another and would offer a salient critique of knowledge claims, regardless of their specialty. The Pepper scheme afforded the possibility of theatrical distance, as it were: the characters in this drama were not authors but rather vectors of distinct *thought styles* whose appeal has varied through different moments in Western social history.

Yet it was not the cognitive claims, theories of truth, or purely epistemological interest that held appeal in this approach. More appealing were the metaphors themselves, in all their heuristic potential. For a metaphor appeals to emotion, aesthetic sense, memory, and will, quite as much as it does to intellect. *Mosaic, mechanism, organism,* and *arena* project four distinct world-views with distinct relations to lived experience (Shibles 1971; Ricoeur 1975). These accounts also reveal that fundamental preferences of cognitive style are rarely explainable in exclusively cerebral terms. Hayden White concluded from his research on nineteenth-century historians that "the best grounds for choosing one perspective rather than another are ultimately aesthetic or moral rather than epistemological" (1973, xii, 426–34). Acceptance or rejection of a particular paradigm, model, or method within the discipline of geography has as much to do with the aesthetic, emotional, or moral connotations of a root metaphor as it does with purely epistemological reasoning.

The succession of metaphors within any tradition raises questions about the interplay of internal and external circumstances. Career stories of geographers reveal important clues about their succession and relative appeal. At any moment of disciplinary history, all four root metaphors to be described in Part 2 may be simultaneously co-present, although one or another may appear dominant within particular periods (fig. 4).[15]

In Sweden, the demise of organicism in the early twentieth century and the virtual absence of contextualist approaches in doctoral dissertations may

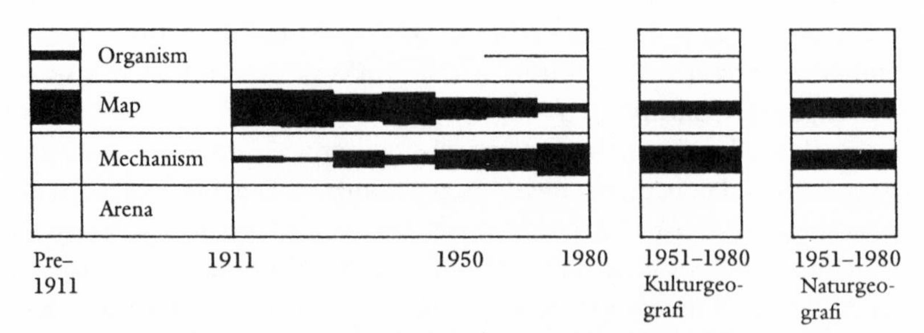

Fig. 4. Dominant metaphors in twentieth-century geography. (Illustration from a study of Swedish doctoral dissertations, 1880–1980.)

set this tradition apart among Western schools (see NL), but there are parallels internationally (Söderquist 1986; Sörlin 1988). The 1920s in America witnessed a concerted attempt to purge organicist prose and to carve out a distinctly American approach relying on inductive and empirical methods. "Sauer distrusted Davis' geomorphology as being built more on deduction than on observation in the field," Leighly observed. "I read none of Miss Semple's writing until later . . . Sauer had known her at Chicago, but already before coming to Berkeley he had discarded her mechanical interpretation of 'geographic influences'" (Leighly in PG, 83). A whole generation of field surveyors, mapmakers, and chorological enthusiasts set about erasing traces of environmental determinism or any imported world-views that had anything to do with imperial politics (PG, 62–65). "We emphasized the concrete," Fraser Hart recalled; "the empirical, the pragmatic, and made little effort to recruit those who might be interested in more abstract and theoretical matters. Hartshorne sufficed. *The Nature of Geography* had appeared before the war, and all right-thinking graduate students slept with a copy under their pillows. A few even read parts of it, and quoting it was one of our favorite indoor sports. Hartshorne was certainly our most quoted and least understood author" (Hart 1979, 111).

Then came World War II, enlisting dutiful chorologists of many lands for strategic and intelligence duties. When they returned from service, many became curious about process rather than pattern, function rather than form, and some spoke of systems analysis and operations research as visionary routes toward a future (better) world (PG, 186–95). As late as 1959, Harold Brookfield, pondering conceptual questions while doing field work in New Guinea, remembered: "I read Hartshorne from time to time, and recall one Sunday afternoon . . . muttering to myself that the chorological approach said nothing to me. If this was geography, then I was not a geographer" (Brookfield, in Billinge et al. 1984, 28). There were others who were quite sure, however, about the way to go. "Our research led to the consideration of geography as general spatial systems theory (terrestrial scales)," Warntz reported confidently, "as part of the more ambitious attempt to expand upon the initial endeavours of the Social Physics Group which had resulted in a mechanical interpretation of various observed mathematical regularities of social phenomena" (Warntz 1984, 145). Torsten Hägerstrand underlined this phrase in George Lundberg's *Foundations of Sociology* (1939): "The ends of science are the same in all fields, namely, to arrive at verifiable generalizations as to the sequences of events." "Given the cackling in geography this new song sounded lovely in my ears," he frankly admitted (Hägerstrand in PG, 248).

The root metaphors *mosaic* and *mechanism* have apparently held strong appeal for individuals concerned about analytical objectivity and empirically grounded propositions. Despite fundamental differences, they offered complementary rather than competing approaches to geography. In fact, one could argue that it was the marriage of these two perspectives and a dowry of mathematical and statistical methods that led to the midcentury reformation and into what Europeans called the "New Geography" and Americans the "Quantitative Revolution." Twin ideas—systems theory and locational analysis—became a burning passion especially among Anglo-Americans, and Richard Chorley and Peter Haggett's writings now replaced Hartshorne's as the most quotable source of orthodoxy (Chorley and Haggett 1967; Gregory 1978). "A society moves under material pressure like a stream of gas," Jacob Bronowski declared in the 1960s, "and on the average, its individuals obey the pressure; but at any instance, any individual may, like an atom of gas, be moving across or against the stream" (cited in Haggett 1965). Reflecting on the transformations of geography in midcentury, Chorley later placed greater emphasis on conceptual rather than methodological aspects of the reformation: "You can't build the core of a discipline around a body of techniques . . . like locational analysis . . . once other disciplines discover spatial auto-correlation, where are you?" Unlike many later enthusiasts for locational analysis, Chorley's own background, with his dual interests in physical and human branches of geography, underlies a firm conviction that there ought to be a core for the discipline as a whole, a core constituted by systems-analytical *logos* (G25).

Members of the succeeding generation, however, would again emerge with distinct metaphorical preferences. While materialist, Marxist, and structuralist critiques tended to use the categories of mechanism (Harvey 1973; Castells 1977; Peet 1977), humanistic and phenomenological prose favored the categories of formism (Ley and Samuels 1978; Seamon 1980). The striking difference between the pre- and postwar generations was that geographers were now being socialized in one or the other branch of the field, their nearest reference group of significance might be in another discipline, and vocational choices toward *poesis* or *paideia* became far less attractive than *logos* or *ergon*.

Of the various symbolic modes used to describe life journeys, the most heuristic is that of a tree (fig. 5). The arboreal representation of a career journey can reveal latent and manifest consistencies as they unfold throughout an entire career.[16] It can also dramatize radically different insights derived from inquiry frameworks suggested by the four root metaphors.

From family and childhood milieus, for example, a young person may receive the impression that academic life is attractive because it cultivates

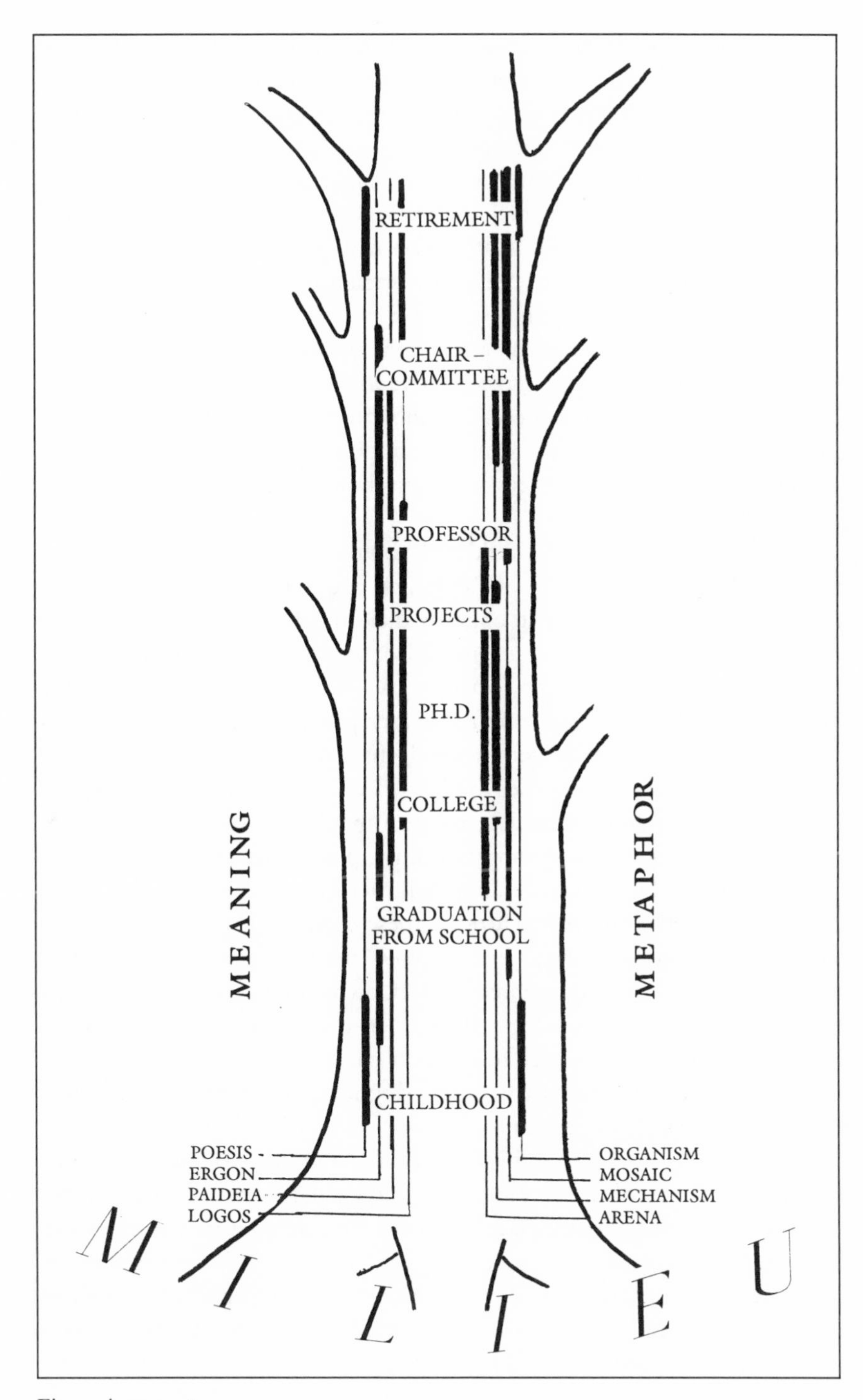

Fig. 5. A career tree

and rewards scholarly virtues. Subsequently, teachers, peers, and external events or opportunities may prompt this person to undertake applied work, the scholarly part being postponed until a later occasion. Again, a charismatic professor or work partner, a book or seminar, may offer compelling arguments for a "systems" or mechanistic conception of reality, so considerable teaching or research energy may be invested in implementing this perspective. Later in life that person may revert to other metaphors, other conceptions of reality which lay dormant during his active years, or may discover new ones. Obviously, it has been at times of impasse, conflict, or passionate rebellion against concurrent orthodoxies that many geographers have hit upon their most creative ideas. For some it has been the prospect of building the sense of nationhood; for others, the opportunity to escape from any and all pragmatic responsibilities, to wander in the woods or circumnavigate the earth, so that thoughts could emerge independently.

An organicist approach would seek to unmask some coherence within the overall evolution of an author's thought and practice, highlighting moments of integration within an ongoing dialectic of opposing forces, rather than documenting precise locations and dates of specific occurrences.[17] A contextualist would prefer to focus on particular career events that were obviously significant for the author, and then proceed to unravel the strands and textures of the context within which this event took place, explaining, as comprehensively as possible, the import of each event for the author's career.[18] Formists and mechanists, on the other hand, would show greater concern for precise documentation of times, places, and circumstances of a career journey.[19] Disciplinary training also tends to shape one's way of interpreting career journeys.[20] Through these exercises it became clear that students with leanings toward *logos* or *ergon* enjoyed the challenge involved in graphic presentation and representation, preferring this to verbal or written forms. For those who seemed eager to dramatize the uniqueness of individual experience, or the creativity of particular places, such exercises seemed anathema. Instead, they preferred to focus on specific career events or on the prose style of particular authors.

Vicissitudes of language, history, and institutional structures have obviously shaped the forest in which these individual trees have grown. Unlike trees, most geographers have been mobile: the places, events, people, and ideas that they record as significant span a wide range of countries and languages. Their preferences of vocational meaning and root metaphor can certainly not be fully understood until placed within the milieus in which their careers unfolded.

Milieu

> One's life journey may be regarded as a puzzle of "milieu pieces" which fuse into one another and all help to elucidate the last piece, the one which is at hand. (Jaatinen in CC, 65)

"As a geographer," Stig Jaatinen wrote, "one feels like a fish on dry land if one does not have the opportunity to experience places and nature with one's senses." It was the Åland Archipelago that drew him most compellingly: "The archipelago is a landscape which can 'get into one's blood' in the same way as Lapland does" (CC, 65). How starkly this contrasts with Max Weber's claim: "The city and it alone has brought forth the phenomena in the history of art. . . . So also the city has produced science in the modern sense" (Weber [1919–21] 1958, 234). "A man's village is his peace of mind," Anwar Sadat reflected; "Once a man's door is shut behind him, no one could enjoy more peace of mind." Facing the challenge of a peace treaty with Israel in the late 1970s, he relied on the "inner strength . . . developed in Cell 54 of Cairo Central Prison" (Sadat 1977, 303–5). Associations between creativity and milieu are myriad, from childhood to old age. Gaston Bachelard wrote on the poetics of space and the potential value of *topoanalysis* (Bachelard [1958] 1964; Tuan 1982). At critical phases in the creative process, especially at initial and synthetic phases, the sensory and emotional aspects of milieu take on special meaning (CC, 119–42).

When senior colleagues are asked today about their reasons for choosing geography as a professional career, most turn to the actual features of their childhood milieus. It was the result of direct contact with the terrain itself and from explorations into its cultural history, or of emotional involvement in the development of nationhood, that many developed their geographical sense of reality (Vilà Valentì, G37; Kostrowicki, G59; Hooson, G60). William William-Olsson recalls his school days at Lundsberg: "The most positive experience [at Lundsberg] was the Värmlandian forest and lake landscape, its Nordic ski winters and the northern lights, and its genuine population of small farmers, charcoal burners, and iron-mill workers" (William-Olsson in PG, 156).

Memories of childhood milieus take different forms. Some writers are visualizers; others recall sounds, smell, and touch. Few emphasize purely intellectual stimuli or curiosities from their childhood days which were not associated with sensory and emotional experiences. "The remembrance of past scenes is inseparable from colour," William Mead remarked, "and disharmony in colour also distracts." It is perhaps natural that a visualizer should be drawn to the open spaces and high plains of the American West, to the wide fells of western Norway, and to the high seas. "Contrastingly, caves and aircraft cabins are disagreeable and I am happier on a horse than in the pressed steel box of an automobile" (Mead in PG, 58).

Choices of vocational meaning and root metaphor have no doubt been affected by the everyday lifestyle of one's childhood days—for example, by one's parents' occupation. Many geographers, it seems, were children of professionals, teachers, civil servants, businessmen, or diplomats; geography for them meant exploration of the *terrae incognitae* or educational service to humanity. There have been "insiders" such as Henri Enjalbert, third child of a farming family in Ségala, who "knew how to use a sickle to harvest and how to adjust a combine harvester. I have worked with hand plough as well as with gang plough tractors. During my trips to Mexico I have always been able to chat with the farmers because I knew what maize was and the soil where it can grow" (Enjalbert in PG, 127). Favored research foci, too, have been influenced by environmental experience. "One's choice of a research subject can come like a bolt of lightning, especially if one feels comfortable in the work environment; a man like Blanchard experienced this in the Alps as well as in Quebec. To work in a setting where one was not in congenial contact with the inhabitants turned out to be rather disappointing; the best guarantee for the pursuit of a chosen work is the fond attachment one feels for the study object" (George in PG, 124). For Karl Butzer, the initial attraction of Spain as focus for archaeological and ecological research had much to do with color and the Mediterranean climate but also was strongly influenced by the social character of the working situation (Butzer, G63). The Mediterranean world has indeed held enormous appeal for geographers from northern and central Europe, even for those who never visited it (Hard 1973; Claval 1986).

Reaching beyond the level of individual experience, however, what of those "structures that permeate individuals . . . but extend beyond their lives" (Dilthey [1913] 1967)? What has been the (structural) "content, value, and purpose" of geography as a discipline in those milieus where it has been practiced? Both Horacio Capel (1981) and Olavi Granö (1981), for instance, have outlined ways in which geography's institutionalization and practical orientation have been influenced by external goals set by society and by the general responses of science to those challenges. Could one identify certain basic human interests that were served, explicitly or implicitly, in the conduct of geographic inquiry throughout the twentieth century?

The establishment of geography as an academic discipline was a deliberate political act, an objectively recorded event in each of our national traditions. From that moment on, one might reasonably assume, practitioners of the discipline would be expected to contribute to the continuing educational and research interests of their respective nations. One avenue of inquiry into *milieu,* thus, was to explore those clusters of public interest toward which geographic discourse seems to have addressed itself. This does not deny the value of "pure scholarship" pursued independently by individuals, many of whom had a global poetic vision. Nor does it deny the potential uniqueness in the formation of an individual's practice. It is rather in the interests of cross-cultural comparisons that this approach sought to uncover traces of external challenges in the practice of geography internationally (fig. 6).

Four constellations of public interest to which geographic discourse addressed itself within our various traditions could be discerned: *identity* (national, regional, or local), *order* (spatial, structural, or administrative), *niche* (resources, *Raum,* or demographic base or potential), and *inventory* (information and communication of knowledge about one's world). The actual connotations of these terms vary between cultures and periods, but

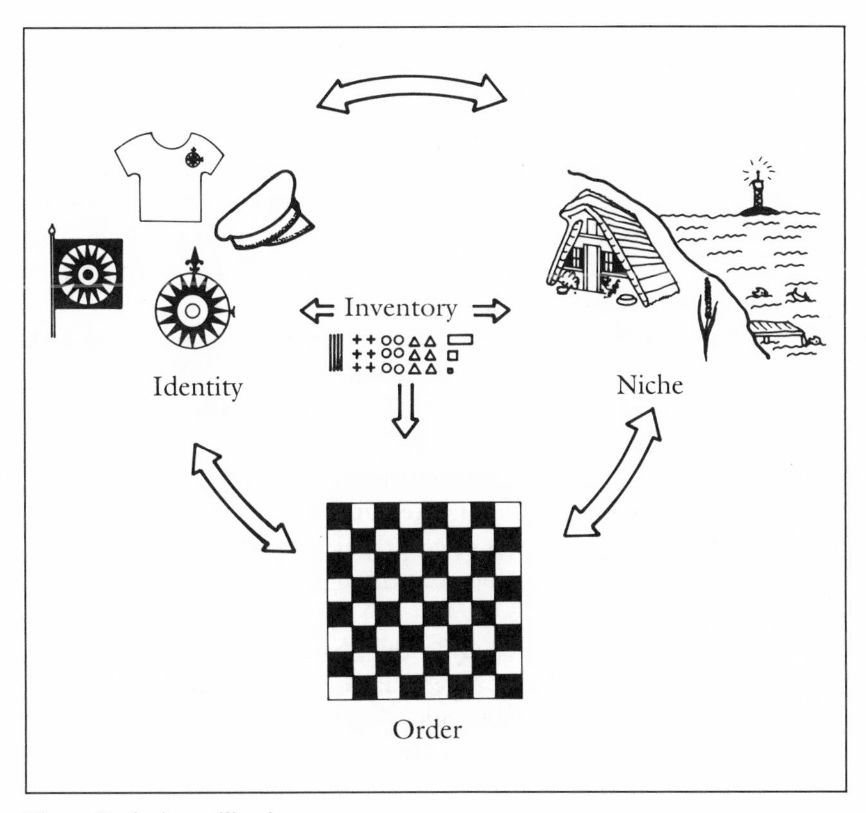

Fig. 6. Enduring milieu interests

the fundamental human interests involved may, in fact, be universal (Cassirer 1946, 1955; Habermas 1968; Lefèbvre 1974; Berque 1982).

Identity subsumes the perennial interest that humans express in developing symbolic and cognitive modes of self-identification and an image of their place in the world. In geography texts, such interests are promoted in early regional works, especially those that highlighted the "personality" of regions, the imprint of civilization on landscape, *Heimatkunde* (home area studies), the national atlas, and later in studies on the sense of place, images, and perceptions.

Order subsumes interests in the management and "housekeeping" of collective life—spatial, temporal, and societal. Such interests are addressed in studies of transportation, urban structures and land use, boundary demarcation and the administrative regionalization of territory, population distribution, habitat, and central place hierarchies.

Niche denotes the interests in bioecological subsistence or expansion, defining territory and reach. Such interests were no doubt explicit in those inventories of natural resources, climates, and soils conducted with a view to improving agriculture, forestry, and economic growth generally; more recent studies of territoriality and proxemics reiterate a longstanding precedent in geography.

Inventory refers to the human interest in information about one's own or others' land. Explorers, atlas makers, and guidebook writers provided "intelligence" and scientifically based knowledge about the surface of the earth and its people. Today the architects of geo-information systems face a similar challenge: to render understandable and reliable inventories of places, resources, events, and processes.

Until midcentury, many major geography texts tried to combine all such interests, as, for example, in classical regional studies. With the rise of thematic specialties and the associated efforts to move from description to explanation, one or the other usually came more sharply into focus. Changes in priority of interest are also evident, following changes of audience and sponsorship throughout the century. The growing dominance of nation-states as primary sponsors of geographic research has also fostered a tendency to address and filter questions through the values and criteria of national interest (fig. 7).[21]

One major value espoused by early twentieth-century "home area study," the first building block in geography training, was to promote a sense of patriotism and regional identity. *Heimatkunde* in northern Europe likewise taught about the connections between identity and niche, *Lebensraum* being at that time still innocent of its eventual geopolitical consequences (E. Kant 1934; Buttimer 1987). Geographers' role in Spain's

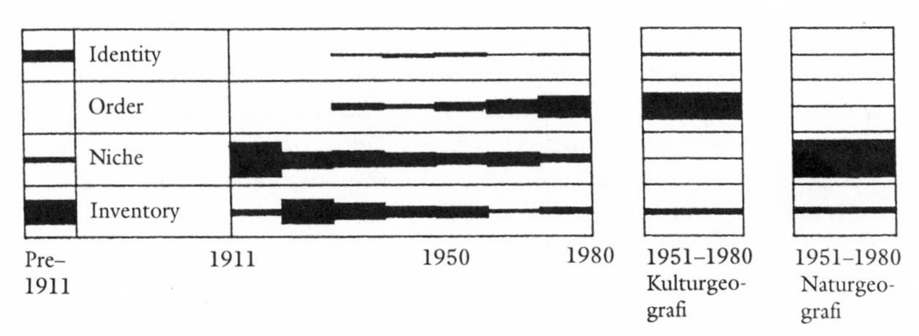

Fig. 7. Geography and national interests. (Illustration from a study of Swedish doctoral dissertations, 1880–1980.)

regeneracionismo was explicitly directed at rebuilding the spirit of nationhood (Gomez-Mendoza and Cantero 1992). The classical French tradition of regional monographs can scarcely be appreciated without setting them in the context of local and regional life in France during the first half of the century. And the connections between *identity* and *niche* were certainly evoked in other schools as well during the first decades of the century:

> For one likes to know what it is one ought to care for. And as one has got to know it, then it has usually grown in value, it has received a richer content and greater importance for oneself. Thus the increased knowledge of the home area will strengthen the feeling for it and make it warmer and richer. But the increased knowledge will also widen the eyes and let the home area emerge as the small part in the big whole, in the fatherland. Then the love of one's home area can grow to include all our land and people. (Nelson 1913, trans. T. Hägerstrand)

With the demise of classical regional description as the discipline's central goal, and with the separation of physical and human branches, both *identity* and *niche* began to take second place to the interests of *order*. Besides, two major world wars undermined national images of human identity and also drew many away from exclusively pedagogical or literary pursuits and toward practical involvement in applied work. Questions of spatial order and functional organization assumed far more importance, not only for the scientific challenges involved, but also for a new vocational orientation for the geographer:

> The thirteen years I spent at Lille marked a burgeoning sensitivity to economic and social issues which has motivated much subsequent research. Max Sorre had successfully planted the seminal ideas, Lille provided the fertile and highly variegated soil in which they could flower: the traditions and prejudices of the locally familiar "patronat," the harsh lifestyles of miners and women in textile factories, a proud pioneering

> region of the nineteenth century confronted suddenly with the demise of coal and fiber-based economy, a ridiculously artificial political boundary severing European people destined for community coexistence, the convergence of immigrants from different periods, diverse races, and competition between Church and State on questions of education—these were my images of a world in transition, not sure of its own future. (Beaujeu-Garnier in PG, 143)

Regional development, urban reconstruction, rationalization of land use patterns, revision of administrative boundaries: the great dreams of the early postwar generation addressed the human interest of *order* throughout Euro-America. The spirit of regional renewal and functional rationality in spatial organization became a common bond among researchers from many different language and cultural backgrounds. In North America, the new esprit would quickly ignite a reform in the practice of geography. "The message was almost a Messianic one," Richard Morrill recalled, "that we needed to assault the citadels of Midwestern tradition, to go out to convert the heathen" (Morrill in PG, 201). In this overriding enthusiasm for rational *order,* few paused to reflect on how these new structures and plans would relate to the other human interests in *identity* and *niche*. Issues of national and regional identity were downplayed after World War II, and issues of niche would lie dormant until the late sixties wave of ecological concern. By then the separation of physical and human branches of the field had become an institutional reality in many schools, and few felt equipped to address questions of order versus niche, of economic versus ecological order.

Other currents of critique began to undermine geography's overwhelming concern about order. Whose interests were really being served: those of managerial authorities or those of "people"? Those of international capital or those of labor? Humanists sought to reveal the consequences for everyday life of ambitious rationalization plans, the pride and glory of a preceding generation's utopian dreams, while materialists charged the same generation with prostitution to power interests, again in the language of order. Perception studies and a heightened awareness of values throughout the 1970s reiterated the challenges facing geography and human interests.

Milieu, examined from the vantage point of human interests, then, opens up a vast range of potential inquiry and reflections about the practice of geography. It remains now to note ways in which the three interpretive themes interweave: how different metaphors or cognitive styles relate to categories of vocational meaning, and how both relate to enduring human (milieu) interests.

Horizon and Integration

The trilogy of themes—meaning, metaphor, milieu—offers a contextual understanding of disciplinary ideas and practices and a heuristic framework for diverse empirical investigations. It also offers bases for critical discussion among researchers from different areas of research specialization. What insight might it deliver on the general issue of integrating knowledge? It certainly helps to elucidate interactions among practical, technical, and speculative knowledge interests at particular moments or periods. Still it remains a framework that serves best to elucidate the synchronic gaze. A critical element is missing from the story, namely, that of scale horizons in space and time. The spatial horizons of research curiosity discerned in Swedish doctoral dissertations revealed the following pattern (fig. 8):[22]

Though there are important differences between the physical and human branches of the discipline, the growing importance of the national scale from World War II on is perhaps characteristic of most Western schools. There are notable differences in the scope of inquiry entertained by the four root metaphors, and also differences in their perspectives on time—past, present, and future.

There does seem to be some consistency, for instance, in the metaphorical styles employed by individuals depending on their vocational commitments: *organism* and *arena* served well as the metaphorical style in which to practice the vocational skills of *poesis* and *paideia*. *Mechanism* and *mosaic* served well for *logos* and *ergon*. There also seemed to be a consistent pattern in the ways different root metaphors readily lent themselves to one or another set of environmental interests: organicist prose was certainly preferred in treatises on national or regional *identity*, and mechanistic or formist models were preferred in studies of *order*. It is conceivable, as Warntz and others

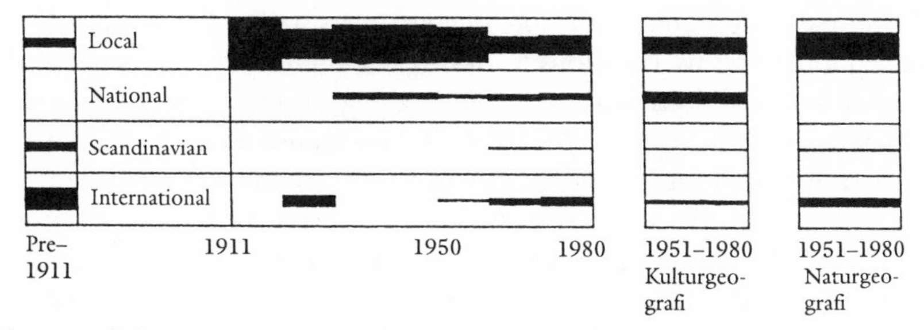

Fig. 8. Scale horizons of geographic inquiry. (Illustration from a study of Swedish doctoral dissertations, 1880–1980.)

have suggested, that integrated frameworks, such as those of *organism* and *mechanism,* do tend to embrace wider (international, global) horizons of interest, while the dispersed views of *mosaic* and *arena* tend to focus on the local and particular (Warntz 1964). Exceptions to such generalizations abound. *Organism,* for instance, inspired synthetic accounts of local as well as of cosmopolitan scale: one could be construed as microcosm of the other. Also *mosaic,* root metaphor of a formistic approach, produced national atlases—surely the geographer's most notable contribution toward clarifying questions of regional and political identity. To grasp issues of integrating knowledge, one also needs to understand how changes in any one of these domains of thought and practice reverberate through the others, regardless of whether the catalysts for change have come from so-called internal or external circumstances.

Even structuralist questions about the interplay of internal and external factors in shaping research practice need to consider issues of scale. In the career accounts studied, recurrent queries arose about the extent to which research styles were products of ingrained habits of thought or "research paradigms" or solely a function of substantive challenge and its material context (Fleck 1935; Kuhn 1970; Teich and Young 1973; N. Smith 1979). There is evidence from geography, as well as from other fields, that once research models have become crystallized around the resolution of particular puzzles and become established as "normal science," they then seek further domains for application. Studies conducted within the framework of *mosaic* served geography well for solving puzzles about spatial *order*. Once its analytical techniques had been refined in the contexts of population distribution, resources, settlement types, or the localization of administrative boundaries, roads, schools, or industries, it was then often blithely applied to other questions, such as regional *identity* and sense of place. Under the etiquette of "sequent occupance," Derwent Whittlesey could extend this metaphor to studies of landscape change and transformation over time (Whittlesey 1929). During the 1960s geographers spoke of "mechanisms" of perception, in terms analogous to those used in previous studies about the mechanisms of industrial location and retailing. It seems indeed that a root metaphor, once in vogue and its analytical procedures refined, continues to exercise appeal, regardless of the problem area chosen for investigation.

It is remarkable that researchers who undergo occasional metamorphoses of ideological orientation often cling to their former root metaphors. There are writers who, as objective "outsiders," waxed eloquently on the cultural landscape, and later, as would-be "insiders," wrote about landscape tastes and environmental perceptions. Some well-established authorities on "systems of industrial organization" later voiced homilies on "systems of domination" or "cycles of poverty/domination." The categorical approaches of formism were suitable for epistemological arguments about "lived" versus "representational" space, whereas those of mechanism were preferred in discourse about revolutions and transformations of the discipline.

But, the materialist might argue, it is not a question of *ideas*. Rather, it is a question of technological developments and their impact on the geographical experiences of humanity as well as on disciplinary practices. The chorological practices and exhaustive regional inventories of pre–World War II days failed to provide precise information on how either side should prepare strategies of defense or attack. Innovations in operations research developed under the shadow of war, for example, and airborne "field" excursions, surely wrought those fundamental changes in map projections, world-views, and eventually root metaphor among many an enlisted geographer, as well as among bureaucrats in the research sponsorship arms of national ministries in the postwar era (PG, 185–95).

Enthusiasts for systems thinking were numerous in the geography schools of both vanquished and conquering nations after World War II. On the root metaphor of *mechanism,* one could anchor not only an alternative theory of truth and ingenious new ways of operationalizing the analysis of social problems but also seeds of a "revolutionary" approach to the challenges facing humanity and the world in the wake of hitherto unimaginable holocaust. For many young geographers in the late 1940s and early 1950s, vocational meaning was found with *logos* or *ergon,* practices confidently imbued with Mannheimian notions of a scientifically objective intelligentsia immune from prostitution to either ideology or utopia (Mannheim 1946).

The contextual turn, already evident in the 1960s and early 1970s, brought critical reexamination of virtually all the inherited thought styles and practices within the discipline. Geography's record in Anglo-American lands showed a marked preference for metaphors of form and process (*mosaic* and *mechanism*): analytically oriented styles of inquiry, be they empiricist or rationalist in rhetorical style, "pure" or "applied" in alleged orientation, populist or managerial in ideological commitment. Might this pattern reflect the discipline's own record of success in claiming *identity, order,* and *niche* within the shifting priorities of funding by national academies or of managerial authorities? Hermeneutic circle, indeed: styles of thought and practice (*genres de pensée*) clearly reflect the lifeways (*genres de vie*) and world-views of their sponsors and audience.

Horizons of scale, in space as well as in time, can scarcely be described in objective Cartesian measures. Cultural values and ideological commitments paint these horizons in emotionally laden ways. The labeling of research themes and the patenting of models and subfields had been achieved by geographers of the imperial nations—those most prolific in the production of geographic texts. Yet it is not at all clear that geographers in other nations, pre- or postcolonial, independent or allied, those with ancient cultural traditions or those busily creating new ones, deviate that remarkably from trends set by the formerly imperial nations. The great challenge of many a geographer in the late twentieth century is how to weigh the interests of disciplinary orthodoxy (and international visibility) against the concurrent interests of a society on the march toward cultural and economic development.[23]

Meaning, metaphor, milieu, horizon: the themes fan out into a complex array of questions for further inquiry. Changing external circumstances and internal discoveries within any of these reverberate throughout the others and vary also depending on horizons of concern. All four interpretive themes point inward and outward: toward the career trajectories of knowing subjects and toward the practice of geography as a product within the political economy of national resources. Janus's concern, if Gunnar Olsson is right, is about creativity (Koestler 1978; Olsson 1984). This framework allows one to examine the correspondences, apparent within particular texts, of vocational meaning, cognitive style, and environmental relevance. It invites synchronic perspectives on particular periods or products. It cannot, *in se,* offer clues to those transitions and breaks in career journeys, those sudden changes of course, which stem from macrosocietal changes in value horizons. To understand the practice of geography, one needs to position it within the wider frame of Western intellectual history. The next chapter addresses this challenge.

Chapter Two

The Drama of Western Humanism

IN SHELLEY'S *Prometheus Unbound*, Asia recites:

> Who reigns? There was the Heaven and Earth at first,
> And Light and Love; then Saturn, from whose throne
> Time fell, an envious shadow: such the state
> Of the earth's primal spirits beneath his sway,
>
> . . . .
>
> Then Prometheus
> Gave wisdom, which is strength, to Jupiter,
> And with this law alone, "Let men be free,"
> Clothed him with the dominion of wide Heaven.
>
> . . . .
>
> He gave man speech, and speech created thought,
> Which is the measure of the universe;
> And Science struck the thrones of earth and heaven,
> Which shook, but fell not;
>
> —Shelley 1820

From Protagoras to Pope, Montaigne to Marcel, Western scholars have claimed that "the proper study of mankind is man." Geographers might qualify this slogan, reminding all that *Homo sapiens* is a terrestrial species, its projects played out in particular environments, even if its diverse cultures have created and been inspired by myths and symbols. Civilizations, Bertrand Russell claimed, can best be understood in terms of their central preoccupations. While China sought to master collective life, he suggested, and India endeavored to master consciousness, the West has displayed an enduring desire to master nature (Nakamura 1980). The conquest of nature has been regarded as the driving motif in Western science and technology, and the Judeo-Christian tradition has often been identified as the ideological source for exploitative attitudes toward nature (L. White 1967; Leiss 1974; see, however, Passmore 1974; Doughty 1981; Kay 1989). Global pronouncements of this kind can unlock major contrasts in cultural interpretations of human nature, but they can also conceal those internal tensions that have existed within each civilization (see Kirk and Raven 1962; Glacken 1967).

It is to the Mediterranean world and to Greek and Roman classics that one turns for insight into the guiding myths and contextual grounding for Western humanism. Not only have Mediterranean models of *humanitas* broadened the scholarly horizons of virtually every European humanist, but the Mediterranean world itself, in all its geodiversity and through all the vicissitudes of its cultural history, has also exercised enormous appeal to those geographers who have today been rebaptized as pioneers of humanist perspectives in the field of geography. Alexander von Humboldt, Carl Ritter, George Perkins Marsh, Vidal de la Blache, Ellen Churchill Semple, Fernand Braudel, Alfred Philippson, Maximilien Sorre, John Kirtland Wright, and Clarence Glacken all drew examples from the Mediterranean world. While speculating on the nature of creation, classical Greek thinkers insisted on seeking insight into the connections apparent in the interactions between mind, nature, and human society. Glacken identified three enduring themes, namely, a designed earth, the influence of the environment on man, and man as a modifier of the environment, which have permeated Western thought on humanity and environment, all tracing their origins to classical Greece (Glacken 1967).

The story of human spirit in the Western world can scarcely be considered without reference to the perennially salient myth of Prometheus, who stole fire from the gods for humanity's sake. In the fifth-century-B.C. play *Prometheus Bound,* Aeschylus portrayed this potential hero who, like Lucifer and Adam, had dared to transgress the rules and was punished. "All this you know, and I'll not speak of," Prometheus replies to the chorus,

> What I did
> For mortals in their misery, hear now. At first
> Mindless, I gave them mind and reason.—What I say
> Is not in censure of mankind, but showing you
> How all my gifts to them were guided by goodwill.—
> In those days they had eyes, but sight was meaningless;
> Heard sounds, but could not listen; all their length of life
> They passed like shapes in dreams, confused and purposeless.
> Of brick-built, sun-warmed houses, or of carpentry,
> They had no notion; lived in holes, like swarms of ants,
> Or deep in sunless caverns; knew no certain way
> To mark off winter, or flowery spring, or fruitful summer;
> Their every act was without knowledge, till I came.
> I taught them to determine when stars rise or set—
> A difficult art. Number, the primary science, I
> Invented for them, and how to set down words in writing—
> The all-remembering skill, mother of many arts.
> I was the first to harness beasts under a yoke
> With trace or saddle as man's slaves, to take man's place
> Under the heaviest burdens; put the horse to the chariot,

> Made him obey the rein, and be an ornament
> To wealth and greatness. No one before me discovered
> The sailor's waggon—flax-winged craft that roam the seas.
> Such tools and skills I found for men: myself, poor wretch,
> Lack even one trick to free me from this agony.
> (Aeschylus, *Prometheus Bound*, 442–71)

The sequels to this play, *Prometheus Unbound* and *Prometheus the Fire-Bearer*, have been lost. Hence perhaps has come the tendency to construe this myth in fatalistic terms: humans hurtling along a linear course of technological mastery of life and landscape toward a reckless end. Others such as Shelley and Goethe have not felt so constrained (Berman 1982). Indeed, as Richard Kearney pointed out, the myth was radically transformed in the light of the New Testament (Kearney 1988).[1] A retrospective glance at the story of geography and humanism suggests another kind of poetic elaboration on the basic myth.

The essential spirit of Western humanism, this chapter argues, could be regarded as the *cri de coeur* (liberation cry) of humanity, voiced at times and places where the integrity of life or thought was in need of affirmation. This quintessentially Promethean élan, in its various phases, is seen to follow a cyclically recurring pattern in Western thought and life symbolized here in the classical figures of Phoenix, Faust, and Narcissus. This trilogy of themes resonates throughout the story of career journeys and in the traditions of professional practices described in Chapter 1: *poesis* (creativity and the nature of humanness, *humanitas*), *paideia* (education and the humanities), *logos* (humanist modes of knowing), and *ergon* (appropriate action; concern for the human condition, or humanitarianism). While Chapter 1 framed a synchronic view of career journeys, Chapter 2 enlarges the horizons to frame a diachronic view of the unfolding of geography and humanism from Greco-Roman times to the turn of the twentieth century.

Phoenix, Faust, Narcissus

Phoenix offers a symbol for those emancipatory moments in Western history when new life emerges from the ashes with prospects for a fresh beginning. In individual careers, as well as in the course of nations, cultural groups, and disciplines, one can identify at least two kinds of emancipatory cry, one seeking freedom *from* oppression, oblivion, or constraining horizons, the other seeking freedom *to* soar toward new heights of understanding, being, and becoming (fig. 9). At times when academy, church, state, syndicate, or proletariat has tried to exercise a monopolizing power over thought and life, a humanist protest has appeared. Socrates told parables to baffle the solipsistic and glib answers of the Sophist virtuosi. Pico della Mirandola challenged ecclesiastical dogmatism in the fifteenth century, paving the way for Giambattista Vico's brilliant recursive interpretations of cultural history two centuries later. Eighteenth-century literati inveighed against the rationalistic claims of the Enlightenment, heralding those masterpieces of emancipatory thought associated with Goethe and the romantics. Kierkegaard, Nietzsche, and Dostoevski in the nineteenth century, Saint-Exupéry, Camus, and Sartre in the twentieth, all pleaded attention for dimensions of humanness that were forgotten or ignored.

Odes to human freedom have certainly not always been evoked by feelings of protest. The sweetest songs in the humanist's repertoire are those that have come gratuitously. Novel ideas in art, literature, science, music, spirituality, or politics have often been expressions of global concern, generous outpourings of a passion for knowledge, life, and beauty. Bruno, Cervantes, Goethe, Shelley, Teilhard, Neruda, and many others have come with messages that were ahead of their times. Characteristically, they were not immediately appreciated, and often they suffered neglect or had to become martyrs before their songs were heard. Phoenix rises gratuitously from the ashes of former dreams; if the climate is inauspicious for receiving a new impulse, it may have to face death through fire before it may reappear.

The key point with Phoenix therefore is its emancipatory message. Its

Fig. 9. Phoenix, *cri de coeur* of humanity, repeatedly reborn to beckon new horizons for life and thought

impassioned cry implies far more than a plea for intellectual freedom or social reform. Humanist movements in Western history have sought to reaffirm moral, aesthetic, and emotional dimensions of humanness. At times they have addressed those dimensions that were repressed by one tyranny or another; at times they have evoked nostalgia for those dimensions that were forgotten or silent; at other times they have pointed toward horizons hitherto unexplored. Success in capturing an audience, however, has often yielded new orthodoxies and structures based on the ardent desire to affirm that which had previously been ignored.

So enters the second symbolic figure: *Faust*. It is surely characteristic of Western ways that, once fresh ideas appear, energies are directed toward the building of structures, institutions, and legal guarantees for their autonomous existence and identity. Goethe's Faust, *eines Menschen Geist in seinem hohen Streben* (the human spirit in its lofty endeavor), stands as central symbol for this phase (fig. 10).

For the pioneering spirit, it is often the idea itself and the new vistas of thought and life heralded by it that are most precious. Details of communication or relevance to ongoing societal interests may be simply tedious. Phoenix therefore welcomes a helping hand in initiating events, and it is often in the bonds growing among the architects of new movements that emancipatory élan is most tangibly felt. And when the idea, along with its

Fig. 10. Faust, "the human spirit in its lofty endeavor," never ceasing to construct for the progress of humanity

legal and institutional bases, has been socially accepted, then a metamorphosis occurs. Pioneering spirits recede, and a later generation directs its energies as much to maintaining and reproducing structures as to sustaining the initial emancipatory ideal. But Faust continues to build—for humanity's sake, inevitably. Otherwise, if he were ever to pause and gaze on his achievement—*Verweile doch! Du bist so schön!* ("Tarry awhile! You are so beautiful!")—Mephistopheles is waiting to steal his soul.[2] Eventually tensions arise between the initial emancipatory ethos and the structures that seek to house and further it. Throughout the many and varied movements of humanistic inspiration, it is possible to differentiate an *ethos,* or fundamental spirit, and a *structure* seeking its incarnation. Socratic seminars became an academy; catacomb communities became a Vatican; networks of mutual aid among workers and peasants turned into syndicated unions and cooperatives; people who felt a common sense of ethnic identity became geographically circumscribed nation-states.

As individuals or groups ponder these tensions and apparent contradictions in their everyday experiences between spirit and letter, dream and reality, a reflective mood sets in which could be symbolized by *Narcissus,* a pilgrim to the muses of Helicon (fig. 11). The poignant evidence is that idealistically conceived plans and innovative movements have become unwieldy structures and self-perpetuating bureaucracies.

On Helicon, Narcissus has two basic choices: to gaze at his own image in the pool of Hippocrene or to hearken to the Muses. As with Phoenix

Fig. 11. Narcissus, bewildered by the contradictions between ideals and reality, reflecting . . .

and Faust, there is an ambivalence. Malaise stemming from tensions between spirit and letter, ethos and structure, may generate critical reflection, archival research, and the quest for clarifying one's identity. Some may insist on interpreting situations through the filters of a cherished self-image; they will emerge with reaffirmation of the status quo. Some may become parricidal, finding all sorts of reasons to condemn their ancestors. The reflective moment could yield insight into all the processes that led to the present malaise, a better understanding of history and the drama of events and their contexts. From this, Narcissus may emerge ready to shed the harness of routine ways and to pave the way for fresh alternatives. In fact, history suggests, one may have to pass through this potentially painful death of former certainties in order to allow the new Phoenix to emerge. History also affords ample evidence of humankind's resilience and its courage to try again.

Such cycles of human experience recur throughout Western history, but there are also threads of linear sequence discernible in the story. None of these cycles ever found humanity or earth in the same condition as before. Through the centuries there have been certain persistent melodies, unresolved paradoxes, which lasted into the twentieth century. Both cyclical and linear accounts help us to place our late twentieth-century challenges within a broader historical frame (fig. 12).[3]

Humanistic Geography

For a few decades now, the term *humanistic geography* has been widely used in geographic literature, and its connotations are many (Tuan 1976; Ley 1981, 1983; Daniels 1985; Rowntree 1986, 1988). What may be subsumed under this rubric varies from one country or language tradition to another, and in some cases the terms *social, cultural,* and *humanistic* have been virtually interchangeable (Racine 1977; Relph 1981; Claval 1984a; Ballasteros 1984; Godlund 1986). For others, "humanistic" has become a prefix for new orientations within such well-established subfields as historical, political, and cultural geography (Daniels 1985; Brunn and Yanarella 1987; Rowntree 1988).

Humanism for some has implied a kind of project to restore human subjectivity to a field where scientific objectivism has become dominant (Ley and Samuels 1978; Mackenzie 1986). Some have emphasized human attitudes and values, others cultural patrimony; some have focused on aesthetics of architecture and landscape, others on the emotional significance of place in human identity (Bowden and Lowenthal 1975; Meinig 1976; Seamon and Mugerauer 1985; Pocock 1981, 1988; Rowntree 1986, 1988). A sub-

Fig. 12. Phoenix, Faust, Narcissus: Cyclical refrain in the Western story

stantial number, too, have advocated human compassion and engagement in the resolution of social or environmental problems—some have offered reminders of twentieth-century pilgrims of peace and of early Cassandra-like warnings of environmental abuse (Thomas 1956; Buchanan 1968; Bunge 1973; Santos 1975; Guelke 1985). There has been widespread concern about humanity and earth, given the shocking record of environmental destruction and the radical transformations in culture and politics (G. White 1985; Johnston and Taylor 1986). From whatever ideological stance it has emerged, the case for humanism has usually been made with the conviction that there must be more to human geography than the *danse macabre* of materialistically motivated robots which, in the opinion of many, was staged by the post–World War II "scientific" reformation (Ferrier, Racine, and Raffestin 1978; Ley and Samuels 1978; Ley 1980; Daniels 1985; Folch-Serra 1989).

Some have looked askance at this humanistic turn as a regressive amnesia, a turning away from practical problems and a retreat into arcane esoterica, or simply as a weak critique of the status quo (Entrikin 1976; N. Smith 1979; MacLaughlin 1986). An earlier generation suspected that humanistic concern might eventually undermine the identity of the discipline, taking the ge- out of geography (Wooldridge and East 1952; Leighly 1955). Today's misgivings reflect not only the doctrinaire antihumanism of structuralism and other philosophical currents but also the internal fragmentation of geography itself. The institutional separation of its physical and human branches, bemoaned today by advocates of environmental sensitivity, was itself the product of an emancipatory project launched by individuals who entered the profession during the interwar period. They sought freedom from the parental bonds of geology and history. Haunted, too, by the ghosts of environmental determinism, they argued for a substantive focus on *space* rather than *environment,* methodological procedures inspired by positivism, and livelier interaction with other scientists. After World War II especially, human geographers proclaimed themselves social scientists; history and the humanities were to become the favored pursuits of only a few. Yet it was from the rediscovery of works by scholars such as Marsh, Vidal de la Blache, Braudel, Wright, and Dardel, and from the heightened awareness of cultural differences in environmental perception, that much of the enthusiasm for a "humanistic" movement emerged in the 1960s and 1970s (Dardel 1952; Lowenthal 1961; Tuan 1976; Relph 1974; Harris 1978).

By the early 1980s, much of the optimism which accompanied this humanistic project had faded. Criticism of humanism came from many quarters: "realists" found the prose of an earlier generation to be quaintly idealistic. Many, disillusioned by the contradictions and tragedies of the Western legacy, condemned humanism as the archculprit, a wellspring of vaunting hubris, a myth overdue for dismissal (Ehrenfeld 1978; Relph 1981). One justification invoked by critics was the conventional practice of identifying the origin of Western humanism with the early fifteenth-century Renaissance in Italy, a movement that also contained the seeds of modernism. Found guilty by association, humanism thus shared the blame for the Promethean excesses of Western humanity. Marxist humanists have, of course, vented spleen on capitalism's perennial ability to produce and reproduce those power structures that have held sway in the global economy (Santos 1975; N. Smith 1979; Harvey G34, 1984; MacLaughlin 1986). But structuralist criticism went farther, eliminating human intentionality and agency from the story. Postmodernists have constructed scenarios of texts reproducing texts in a labyrinth of self-reflecting mirrors in a manner precluding any assumptions about human intentionality or meaning in lived reality (Jameson 1983; Said 1983; Kearney 1988). A sense of imprisonment in one's own cultural world, of claustrophobia and nihilism, of "hitting one's head against the ceiling of language" (Olsson 1979; Dematteis 1985) characterized some of the critical thought within geography during the 1980s. One became aware of the many ways in which the entire intellectual heritage of the West had mirrored and been mirrored in the peculiar social history of Western humanity.

Even over this relatively short stretch of disciplinary history, one could observe a cyclical movement in intellectual life. In retrospect, the 1960s were a Phoenix time for many, with new life bursting out all around. By decade's end and throughout the 1970s, one witnessed a Faustian will to create new subdisciplines, societies, and specialty groups. The late 1970s and early 1980s revealed many contradictions between ethos and structure, and a critically reflective mood set in. Some became nostalgic for the past; others reaffirmed the status quo, and some envisioned prospects for a new dawn (Rowntree 1988; Ley 1989).

Mediterranean Musings

The legacy of Greek thought is as multifaceted as it is enduring. In pre-Socratic times, it is alleged, ontological speculations embraced humanity and earth in an integrated way. *Physis* (the whole of physical and living reality) and *Nous* (mind, reason) were integral parts of the world-views of Heraclitus and Empedocles (Kirk and Raven 1962). Plato is generally credited with that fundamental distinction and the eventual separation between mind and matter, thought and being, metaphysics and physics, which have since characterized Western approaches to knowledge. To his school also is ascribed the glorification of intellect among human qualities and the enthronement of humankind as superior to all other life forms on the earth. In that famous Socratic tradition, however, there were essential differences between Aristotelian and Platonic world-views (see Chapter 3). Was there a Supreme Artisan with a plan for the wise workings of nature, or was the world a theater of becoming, of nature bearing within itself the seeds of ongoing creation? Through centuries of Judeo-Christian and Arab philosophy and geography, definitions of human nature still echoed the basic tenets of classical Greek thought.

All brands of humanism have not insisted on a return to the classics. The romantics of early nineteenth-century Germany and New England, while deeply impressed by classical lore, also found a universe of inspiration in their immediate surroundings (Bunksé 1981; Kohak 1984). Marxist humanism and Sartrean existentialism found little need to ground their appeal in classical sources. Christian humanism has focused on the human soul's quest for eternal salvation, the Greeks figuring as one among many edifying cultures which articulated virtue, compassion, and love of truth. The literature about humanist approaches to knowledge (*logos*) particularly reveals the love-hate relationship to Greek paragons: Plato and Aristotle especially have been successively enthroned and impeached down through the centuries.

The term *humanism,* it has been claimed, is of Roman origin (Heidegger 1947). For Cicero, *humanus,* as opposed to *barbarus,* denoted the civilized Roman citizen, erudite in literature and culture and ready to assume a responsible role in civic life. *Humanitas,* for better or for worse, became indelibly identified with *Romanitas,* a civility won via Greek (especially late Hellenic) education or *paideia* (Heidegger 1947). In stark contrast to this stance was the slogan of Terentius: *Homo sum: nihil humani a me alienum puto* ("I am a man: I regard nothing that is human as alien to me"), an appeal to universalism which Saint Augustine, among others, made central in his definition of humanism. Throughout all subsequent flowerings of humanism in Europe, until the eighteenth century, there was a tendency to define humanity in terms that would differentiate it from inhumanity or barbarism and an opposite appeal to the *uomo universale,* capable of becoming a civilized participant in universal humanity.

Characteristic, too, has been the proposal that the achievement of *humanitas* would require a *studium humanitatis,* a schooling in philosophy, literature, rhetoric, and the arts, preferably based on Greek models. Humanists have often proclaimed that education involves more than indoctrination; many have favored *Bildung* (*paideia*), which encourages a sense of responsibility for self-development and humane behavior. Some have placed their strongest emphasis on *poesis,* the art of evoking curiosity, critical reflection, and invention. As for individual and social action and behavior (*ergon*), debates about the nature of human communities, democracy, civil rights, and moral freedom have also consistently found exemplars in Greek and Roman ancestry. In terms of humanity's modes of inhabiting the earth, Rome has also bequeathed the tension between "arcadian" (e.g., Vergil's *Eclogues*) and "imperialist" (e.g., Seneca or Cicero's *Letters*) landscape models, a tension documented in recent writings about humanity and nature (Pepper 1984).

Few Mediterranean legacies could rival that of Arab scholars throughout Europe's so-called Dark Ages. While Platonic ideas dominated intellectual life along the northern and western coasts and peninsulas, Aristotelian ideas bore rich fruit along its southern and eastern shores, extending from the heart of Asia to the Atlantic shores of the Iberian Peninsula. Christendom may have slumbered within the framework of a Jerusalem-centered world-view, but Arab sailors, pilgrims, traders, and cartographers filled in details on a vastly more extensive *oecoumene*. Had it not been for the fertile meeting ground of Arab, Jew, and Christian in Cordoba and other Moorish towns up to the twelfth century, geography might never have been part of the Renaissance Phoenix. Yet for many a Western humanist, the Greco-Roman classics have been regarded as far more significant than those of the Moor.

Humanitas (*Poesis*)

Literary historians traditionally identify the fourteenth and fifteenth centuries A.D. and the Florentine Academy as the proximate origin of modern humanism in the West. Art historians single out Petrarch or Fra Angelico; explorers emphasize Henry the Navigator; others pick Dante or Leonardo da Vinci as paragons of humanism. But few could question the Phoenix mood of this period rising from the ashes of medieval times, heralding an age of adventure (de Santillana 1956). The Renaissance, oriented to both past and future, contained the prototypes of medieval security and projected visions of the individual as explorer of new powers and unlimited horizons (Eco 1986). Here was a world in which the emancipatory cry of humanity was about reform in the practice of religion, a reform involving a return to origins in the Sermon on the Mount as well as in a rediscovery of the Greek classics. In 1486, Pico della Mirandola published *An Oration on the Dignity of Man,* a document that proclaimed human nature to be ontologically free and responsible.[4]

The daring deeds of Renaissance scholars in the realms of art and architecture, music and painting, architecture and philosophy, each championed some quality of humanness which had been forgotten, oppressed, or silenced. For *homo ludens,* there were the delights, albeit fleeting, of arcadian springtime; for *homo faber,* the marvels of technology, the plastic arts, and freemasonry beckoned; for *zoon politikon,* the egotistical world of princes and potentates was symbolized in magnificent villas. For all of humankind, the image of the human body as microcosm of the universe was an insight that could steer imaginations to explore connections between alchemy and medicine and between geography and astrology (Glacken 1967; Mills 1982). The rediscovery of Arab scholarship with its medical, cartographic, and mathematical genius opened new frontiers for the study of nature. Ptolemy's *Almageste,* recovered only in the thirteenth century by Dominican monks, fired the imaginations of Copernicus, Galileo, and others to construe the world as a "perfect work of art," and to reposition the earth itself in cosmographic and astronomical contexts (Hasr 1964). For *homo geographicus,* the fifteenth century marks one of the most striking moments of creative discovery (Broc 1986). Few foci of attention could rival the *mappae mundi* as windows onto human cultural diversity beyond the known world of Christendom. Between 1409, when Jacobus Angelus translated Ptolemy's *Geographia,* and 1570, when Ortelius's *Theatrum Orbis Terrarum* appeared, the world images of Europeans were radically altered. Spurred on by fabulous tales and gross misconceptions about the size and shape of the earth and its exotic cultures, horizons of space and distance were pushed back by the daring voyages of Portuguese, Italian, French, Spanish, and Dutch explorers. The finite and circumscribed world of medieval orthodoxy gave way to vast expanses for exploration, conquest, or conversion (fig. 13).

Within the span of one and a half centuries, a dramatic range of luminaries made their mark. Leonardo da Vinci pioneered on the frontiers between art, technology, and philosophy; Michelangelo and Fra Angelico bequeathed treasures of painting and sculpture. Prospects for easier communication among scholars were opened up via the invention of print, eventually liberating discourse into vernaculars other than Latin. A reformation in the Christian church promised deliverance from Vatican hegemony.

The rediscovery and dissemination of classical works were certainly due to the expansion of curiosity about the world among Italian humanists. It was surely from these texts, despite all their misconceptions about world geography, that the voyages of Columbus and Cartier drew their inspiration (Broc 1986). The further awakening of geographical curiosity may well have been due to the mobility of people and ideas which subsequent voyages, commerce, and navigation occasioned and the development of new hori-

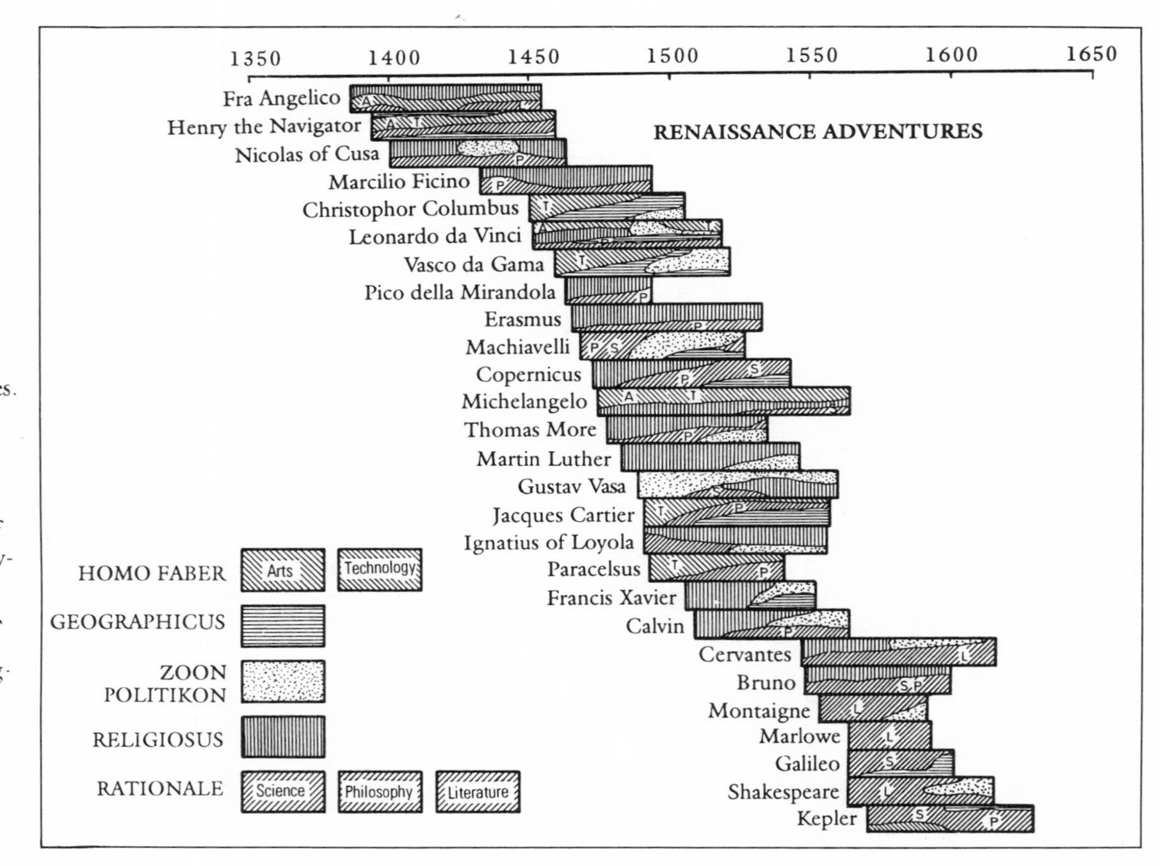

Fig. 13. Renaissance adventures. An emancipatory moment for artisans, explorers, and active participants in political, religious, and intellectual life. Note the pluralistic interest of early Phoenix voices, the growing specialization of Faustian constructions and their critics, and the reemergence of pluralism in the "twilight longings" of Narcissus before the dawn of a new Phoenix.

zons for *homo faber, homo viator,* and *homo oeconomicus* (Braudel 1966; Broc 1986). Travel accounts were quickly translated into other vernaculars, stimulating geographical curiosity among people who, unlike academic humanists, were not well versed in Latin (Matos 1960). The Renaissance was "a quick rush along the whole front, urged on by staggering innovations which could not be grasped in their full import, a surge which in less than three centuries loses itself in the unending rapids of the scientific era" (de Santillana 1956, 9).

It was a time of *poesis,* leading to wider horizons for cosmology, cartography, and ethnography, and it also witnessed remarkable developments in landscape design and capitalist enterprise (Cosgrove 1984). In virtually all respects, it must have presented subversive challenges for the Faustian structures of the day. New knowledge about *terrae incognitae* challenged orthodoxy and undermined the power of established authorities, most especially ecclesiastical ones. During the fifteenth, sixteenth, and seventeenth centuries, emancipatory appeals of humanism were addressed to questions of *homo religiosus;* the primary concern was to reconcile man and God, man and world, a world whose contours were no longer accepted as finite. The theme song of the humanist was that human persons could work out this reconciliation through their own activities, as responsible and creative agents within a potentially expanding universe.

As Faustian interests clashed on issues other than theological, the emancipatory élan underwent a testing by fire; de Santillana portrays this in his discussion of the drama of late Renaissance times:

> Bloody religious struggles are leading up to the climax and holocaust of the Thirty Years' War, from which Europe was never morally to recover, a period in which theologians are as busy as a barrel of monkeys sharpening knives in Councils and Synods, in anathema and counter-anathema, there is a wonderfully positive content, and a content of evangelical hope, in that fight for justice and tolerance, in that affirmation of peaceful and cosmic religiosity, which goes from Erasmus all the way to Voltaire. (30)

Antihumanist protest also came from ecclesiastical sources. The horrors of the Inquisition in Spain and Italy are well known, but some of the most explicit condemnations of humanism came from the leaders of the Reformation, Luther and Calvin, who flatly denounced Erasmus.[5] The crucial stumbling block was the doctrine of original sin, anathema to the humanist. One can therefore appreciate why ardent pioneers of scientific humanism of the eighteenth century, such as d'Alembert, de Lamettrie, and Diderot, felt so strongly about the need to free human minds from clerical oppression. In this case, as in so many previous and subsequent ones, the oppressor's mantle became discernible in the credo and structures that followed. The *Encyclopédie* appeared as a secular catechism, a believer's compendium of facts about the true knowledge of reality. Later, Auguste Comte (1798–1857) announced his prospectus for positive science and a new society based on castes of specially trained experts, a scenario not that different from those ecclesiastical ones that he had deemed inimical to humanity.

At the evening of the Renaissance period, Narcissus could claim an audience. Montaigne (1553–92) satirized cultural myopia and appealed to critical self-understanding among Western scholars. Appalled by Christian attitudes toward pagans and Mohammedans, he wrote that "each man calls barbarism whatever is not his own practice; for indeed it seems we have no other test of truth and reason than the example and pattern of the opinions and customs of the country we live in. *There* is always the perfect religion, the perfect government, the perfect and accomplished manner in all things" (1: 31, 152).

Together with Nicolas of Cusa, Montaigne stands out as a pioneering advocate of cultural relativism, and also as a critic of the Western world's anthropocentric attitudes toward nature: "Can anything be imagined to be so ridiculous as that this miserable and wretched creature, who is not so much as master of himself, but subject to the injuries of all things, should call himself master and emperor of the world, of which he has not power to know the least part, much less to command it?" (Montaigne 1603, bk. 2, chap. 12).

Marlowe, a contemporary of Galileo, had the chorus sound an ironic note on *Doctor Faustus* (1604);

> His waxen wings did mount above his reach,
> And, melting, Heavens conspir'd his overthrow;
> For, falling to a devilish exercise,
> And glutted [now] with learning's golden gifts,
> He surfeits upon cursed necromancy.
> Nothing so sweet as magic is to him,
> Which he prefers before his chiefest bliss.
> And this the man that in his study sits!
>
> (Marlowe 1604, 206)

"'Tis all in peeces," John Donne claimed in 1616, "all cohaerence gone." And Prospero, in Shakespeare's *The Tempest,* envisioned

> The snow-capp'd towers, the gorgeous palaces
> The solemn temples, the great globe itself
> Yea, all which it inherits, shall dissolve
> And like this insubstantial pageant faded
> Leave not a rack behind.

As Renaissance yielded to Enlightenment and Phoenix to Faust, the foci of scholarly concern became more narrowly specialized, and waves of intellectual creativity swept from the Mediterranean to northern lands (Buttimer 1989b). As previous generations had argued about the human soul's quest for eternal salvation, post-Renaissance debates would focus on rationality, on the intellectual faculties of *Homo sapiens*. In late Renaissance times, too, other human qualities were extolled, albeit not in Aristotelian or metaphysical terms. Francis Bacon (1560–1626) emphasized *homo faber* and laid the foundations for experimental science and engineering. Hobbes's *Leviathan* (1672) drew attention particularly to *zoon politikon,* sowing seeds for a revolutionary change in traditional beliefs about civil rights and responsibilities. The torch of progress, and hence new, compelling definitions of humanity and nature, was borne by the Galilean spirit of scientific inquiry based on testable hypotheses rather than on traditional dogma. It would lead directly to the triumphs of Cartesian and Newtonian science from the seventeenth century on (Koyré 1957). Thereafter ontological questions yielded to epistemological ones: "what" and "wherefore" deferred to the "how," "when," and "where" of humanity and its terrestrial home.

Humanist Modes of Knowing (*Logos*)

The Enlightenment heralded fresh prospects for human reason and life. Réné Descartes's *cogito ergo sum* (I think, therefore I am) was a statement about both human nature and human knowledge (Descartes 1637). Since the seventeenth century, this credo had been a primary target of humanist critique. Giambattista Vico (1668–1744) proposed an alternative approach to science (Vico [1744] 1948; Kunze 1984). Cartesian rationality, he argued, could not really account for human ingenuity as expressed historically in the common-sense language and behavior of diverse cultures. Nor could it explain human judgment in moral affairs (*prudentia*). It also implied that the human faculties of imagination, fantasy, and sense perception were somehow "nonrational," and therefore inadmissible in the conduct of science. In Vico's view, the Cartesian approach thus made it virtually impossible to understand history. The kind of cognitive clarity sought by Descartes was something that could only belong to the Creator with respect to his own creations. As man had not created nature, he could never know it with certainty. He could only know history, which was his own creation. "Philosophers have tried to arrive at knowledge through the realm of nature . . . and they have neglected to reflect on the world of nations, or the historical world, which was created by man" (Vico 1744, 4, par. 331).

In Vico's approach, humanist modes of knowing sought the inventive quality of human persons and groups and their ability to discern connections and relationships among dissimilar things. One should look to this ingenious faculty for insight into metaphorical thought, which for Vico could unlock the secrets of human culture and history (Nicolini and Croce 1911–41:1, par. 183; Mills 1982; Kunze 1984). Vico was affirming the principle that modes of knowing inevitably imply assumptions about the nature of being. Here was a line of thought that would yield a rich harvest in the works of von Herder and later Michelet, and indeed would afford one major source of inspiration to twentieth-century *géographie humaine* in France.

Less sanguine images of human nature were also projected in the English-speaking world of the eighteenth century. Theologians and moralists harangued about humankind's perennially mixed motives, self-deception, and irrationality. Reminiscent of Juvenal and Plautus of Roman times, the theme of *homo hominis lupi* (man as wolf to man) permeated the prose and poetry of the day. Oliver Goldsmith, by no means the most brash of eighteenth-century writers, revealed that pessimistic attitude toward human nature which was characteristic of his times:

> Logicians have but ill defin'd
> As rational the human mind;
> Reason, they say, belongs to man,
> But let 'em prove it if they can
> . . . .
> [I] must in spite of 'em maintain,
> That man and all his ways are vain,
> And that this boasted lord of Nature
> Is both a weak and erring creature.
> That instinct is a surer guide
> Than reason—boasting mortals' pride;
> And that brute beasts are far before 'em.
> *Deus est anima brutorum.*
> (Goldsmith 1887, 194)

The idea that instinct might be a better guide than reason was a theme that would later inspire naturalists and pragmatists in North America. Individuals faced with the artificialities of culture and the complex interplay of passion, emotion, and reason within themselves sought refuge in universals. In the Creator's grand design, Alexander Pope suggested, there was provision for a potentially creative outcome if harmful things were bounced off one another. In his *Essay on Man* (1733–34), a widely read item in eighteenth-century England, he shared a view on the tensions between passion and reason within the human person:

Passions, like elements, tho' born to fight,
Yet, mix'd and soften'd, in His work unite:
These, 'tis enough to temper and employ;
But what composes Man, can Man destroy?
. . . .
Each individual seeks a sev'ral goal,
But Heav'ns great view is one, and that the whole.
That, counterworks each folly and caprice,
That, disappoints th'effect of every vice.
(Pope, Epistle 2: 111, 235)

From a geographic vantage point, the relevant issue was how such models of humanity might elucidate diverse forms of collective living and the politics of terrestrial homemaking. What evidence could be gathered from the *sensus communis* among people of various nomadic, sedentary, or commercial livelihoods? Buffon's *Histoire naturelle* (1749–1804) documented the dependency of livelihood and social organization on environmental conditions; through this he made a strong case for monarchy. But in Montesquieu's *De l'esprit des lois* (1748), for all its environmental determinism, there was again an echo of Pope's recipe for the management of humankind's warring passions. Add this to the doctrines found in the writings of Polybius, Machiavelli, and Jean Bodin, and one had some practical recommendations for the would-be rulers of nations: useful outcomes could emerge from balancing harmful things with one another.

Here was a "humanist" theory not altogether irreconcilable with the cutting-edge "scientific" theory of the day concerning nature and the universe. As Lovejoy and others have suggested, the U.S. Constitution (1787) expressed a view of human nature and political life reflecting the balancing of forces analogous to Newtonian physics (Lovejoy 1961; Warntz 1964). The teaching of geography in colonial colleges during the seventeenth century mirrored this mechanistic *Zeitgeist*. Newton's *Principia,* amply illustrated with Varenius's "general geography," provided the basic text for the required course on geography and astronomy popularly announced as "The Uses of the Globes" (Newton 1687). Many of the architects of the 1787 Constitution had already benefited from education in such curricula (Warntz 1964).

In the wake of the American and French revolutions, Adam Smith's *Wealth of Nations* (1776), and the "discovery" of lands and peoples hitherto unknown to Europeans, the focus of speculation about human nature passed over to the realm of the social. Utopian revolutionary visions of a better world and the progress of humanity emerged from the writings and deeds of scientific humanists (de Dainville 1941; de Lubac 1944). Yet, critical ontological issues remained: was sociality innate among humans, or was it taught or learned? Were there universal traits of the *zoon politikon,* or were all observable traits simply a product of socialization? Were humans naturally disposed toward compassion, generosity, good will toward fellow humans—those virtues extolled in classical humanism—or were they naturally disposed toward competition, conflict, and preying on their fellows? Adam Smith surely was unequivocal on this: the Industrial Revolution would be grounded on profit-seeking individualism. The U.S. Constitution, however, affirmed the Enlightenment credo that the common good could be guaranteed through well-constructed political machinery.

In Europe there were misgivings about such a blunt creed. Voltaire's *Candide* (1759) elegantly satirized the Enlightenment project of rationality in human affairs. From the mechanical certainties and all-embracing explanatory power of Newtonian science and, perhaps more emphatically, from the "dark Satanic mills," romantic writers sought to rescue human nature. As pre-Renaissance scholars had sought emancipation from the well-structured, stable, and closed worlds of medieval times, romanticism sought to liberate humanity from the all-embracing mechanical certainties of the Enlightenment. Schiller, Blake, Milton, Keats, Dryden, and Pope, in satire or grand tragedy, would sing of humanity's nonrational qualities, its passions and desires, its moral and aesthetic senses, and the contradictions between word and action. Nature, too, was far too mysterious to be scrutinized by physical science; it was, in Schelling's words, "the sacred and primary force," a "great chain of being" (Lovejoy 1936). Goethe (1750–1832) bequeathed Faust, personification of the spirit of this age, who pledged to unravel the mysteries of cosmic harmony:

> How all things live and work, and ever blending,
> Weave one vast whole from Being's ample range!
> How powers celestial, rising and descending,
> Their golden buckets ceaseless interchange!
> Their flight on rapture-breathing pinions winging,
> From heaven to earth their genial influence bringing,
> Through the wild sphere their chimes melodious ringing!
>
> A wondrous show! but ah! a show alone!
> Where shall I grasp thee, infinite nature, where?
> Ye breasts, ye fountains of all life, whereon
> Hang heaven and earth, from which the withered heart
> For solace yearns, ye still impart
> Your sweet and fostering tides—where are ye—where?
> Ye gush, and must I languish in despair?
>
> (Goethe, 1808, 26)

For those many geographers whose energies for a century or so had been spent on basic compilations of information and perfection of mapping techniques, Cartesian geometry and Newtonian mechanics offered new frontiers for rational and elegant renderings of the earth's surface (Broc 1974). Many would also hearken to the optimistic promises of eighteenth-century *Encyclopédisme* and the prospects of improving the human condition through societal applications of scientific rationality (de Dainville 1941). Like Pope Alexander IV, who in 1494 drew the Tordecellas line to separate the Spanish and Portuguese territories in the New World, Immanuel Kant would define the legitimate territories of intellectual curiosity for particular fields in the early 1800s (see Chapter 3). Geographers should focus on space, the outer sense; to historians belonged the study of time, the inner sense, and all that this implied in terms of emotion and human experience (Tuan 1978).

It was still in a Kantian spirit, while transcending the letter of his epistemological law, that geography's two great pioneers, Alexander von Humboldt (1769–1859) and Carl Ritter (1779–1859), made their decisive contributions to the discipline. Far from armchair speculation about human nature or worries about boundaries separating science and humanities, Humboldt's *Cosmos* remains even today the unrivaled model for a geography imbued with the humanist spirit (Humboldt 1845–62; Bunksé 1981). Together with Carl Ritter, the other acclaimed father of modern geography, Humboldt moved geography beyond the routine-operational compiling of information, classification, and mapping of earth features. The earth and its panorama of diversified landscapes contained for him the drama of civilization and biosphere. In their actual writings, the old distinctions between Platonic and Aristotelian ontology would reappear: Ritter's *Erdkunde* read the earth's landscapes as script of a divine plan for humanity (Ritter [1815] 1862), Humboldt's *Cosmos* found in the cosmos itself the "sacred force . . . animated by the breath of life" (Humboldt 1849, 3). While attuned to the general quest for *Ganzheiten*, which characterized their day, both defied the Kantian boundaries of geography as mere mapmaking: time, process, causal connections were all included in their mirrors on reality. Humboldt was fascinated by the diverse ways in which humans had internalized nature and landscape. As a pioneering voice in the exploration of environmental perceptions (Bunksé 1981), he would actually subvert the Kantian orthodoxy that geography was simply the description of the earth's surface.

Meanwhile, in literary and philosophical circles of the nineteenth century, a narcissist mood was evident. Queries were raised about the contrast of scientific and humanist modes of knowing, and there was a search for those elements in thought and life that had been forgotten or ignored.

Schopenhauer (1788–1860) once proclaimed "The World is my Idea," emphasizing the fundamental "will to power" underlying Western approaches to truth. Herder, Goethe, Hölderlin, and disciples of Hegel protested not only the perceived threat of Cartesian scientism but also the limits on mankind's historical consciousness. Nietzsche (1844–1900) sought to reintroduce questions about the passions, volition, and aesthetics into the discourse on human nature. Appealing, in characteristically humanist fashion, to classical sources, his *Zarathustra* and *Birth of Tragedy* proclaim the essential tensions between Apollonian and Dionysian elements in Greek drama. As yin to yang in Oriental lore, he claimed, so might the reciprocity of Dionysus and Apollo in Greek literary creation be regarded (de Lubac 1944). Socrates was blamed for the suppression or death of Dionysus and the enthronement of the rational tradition that came to dominate the West via Apollonian styles of thought and life. Kierkegaard (1813–55), albeit a faithful admirer of Socrates, lent weight to Nietzsche's struggle against Hegelian conceptions of universal humanity and the determinism of historical processes within which the individual was swept along as a passive automaton, bereft of heart, soul, or personhood. For Nietzsche, Kierkegaard, Heidegger, Camus, Sartre, and the twentieth-century existentialists, the challenge was to evoke awareness of emotional, volitional, aesthetic, and passionate aspects of human nature, human knowledge, and human action.

Scientists of the late nineteenth and early twentieth centuries, such as Comte, Darwin, Marx, and Freud, dismissed or transcended those antipathies between humanist and scientific modes of knowing and pressed ahead with Faustian zeal into inquiries that eventually shed new light on humanity as a terrestrial species. Each of these bore a potentially emancipatory message in a world now less tolerant of dogma or ethnocentric conceptions pronounced by national, ecclesiastical, or academic authorities of the day. Darwin freed imaginations into a conception of humans as participants in the general processes of natural evolution, as phenomena to be explored via scientific observations rather than culturally ingrained beliefs. For Marx the issue was not one of defining humanity but of liberating it: a "species being" working out its own liberation in the concrete circumstances of life, labor, and history. Freud would later plumb the turbulent depths of the human psyche, opening up unexplored realms of human nature and promising once more a universal definition of human nature. Traditional notions about freedom, about the human individual as author of thought and action, were all challenged.

Such ontological queries about human nature and epistemological queries about modes of knowing captured the attention of the nineteenth-century geographers Reclus and Kropotkin. Far more fascinating and energy consuming for the majority, however, were those fresh perspectives on the dynamics of the earth, its atmosphere and oceans, its climates and biotic life, which were explorable within the framework of natural science. Volumes of new information about cultures and foreign ways of life flowed in as imperial nations sought to colonize and exploit distant realms and educate their schoolchildren in "world geography." The nature of geography's sponsorship and audience had changed radically since the days of Humboldt and Ritter. Foremost among its burning curiosities would be questions of environmental determinism, a subject which, like original sin, no self-respecting humanist would wish to touch.

Doctor Faust now enters the drama in earnest. As nation-states created chairs for disciplined knowledge, and formal criteria were defined by which the domains of science and humanities could be circumscribed, epistemological distinctions incarnated themselves in institutional separations. Functional specialization in university life was assured by Faustian structures whose architecture reflected Cartesian (or Weberian) rationality and whose content and functions eventually fulfilled Comtean dreams of positive science. Geography, precariously perched between the competing knowledge claims of science and humanities, faced an especially challenging situation.

The Humanities (*Paideia*)

A great controversy captured attention at the Cathedral School in Paris during the twelfth-century debate on the functions of a university. The powerful voice of Bernard of Clairvaux argued that the function of university education should be the formation of the whole person, namely, education in moral as well as intellectual virtue; Abelard, master of rhetoric, argued that emphasis should be placed primarily on the intellectual. Abelard lost the battle but won the war. The Western university adopted, de jure and de facto, the principle that intellectual formation was to be its primary purpose. Given this fundamental option, rationality eventually demanded that each field be allotted its appropriate place, its own special agenda, and rules for discourse.

The *humanities* consisted of a range of knowledge fields whose central focus rested on the study of humanity, that is, history, literature, arts, rhetoric, and the classics. Already in fifteenth-century England, William Caxton (1422–91) drew sharp distinctions between "humanities" and "divinities," the former regarding humankind as an object of study in itself, not as part of nature, which was the concern of biology, nor as an object of divine grace, which was the concern of theology, but as a reality in its own right. Where the boundaries were to be drawn with respect to the social sciences was not then a problem, but it would later became a controversial issue. The tradition of liberal education, *paideia,* affirmed the importance of intellectual and moral goals. It sought to train and educate students in certain edifying fields of expertise, for example, literature, rhetoric, and the arts, and also to develop a humane attitude toward one's fellow beings. Both goals were clarified by Cardinal Newman in his *Idea of a University* (Newman 1852).

The specific range of knowledge fields incorporated in faculties of humanities varied from one country to another. The assumption that the cultivation of *humanitas* could be facilitated by a return to Greek and Roman sources was explicit in most programmatic statements about the humanities in the Western university. Platonic distinctions between spirit and matter, parables of Phaedrus and the Platonic Cave, and the heroic challenges of Socrates permeated texts in moral and natural philosophy. The Ciceronian antinomy of *humanus* and *barbarus* served not only to motivate students but also to bolster cultural and national identity among the potential citizens of expansionary empires. The Terentian-Augustinian slogan, *nihil humani a me alienum puto,* also served as motto for all who aspired to liberalism, pluralism, and international brotherhood.

The institutional framework, however, demanded anything but an ecumenical attitude toward knowledge. Each field should cultivate its own ground, perfect its own skills, and affirm its own identity within the Faustian program for progress which one nation after another launched after the Enlightenment. Thus the humanities, especially in the nineteenth century, would be defined in terms distinguishing them categorically from the natural sciences. *Verstehen* (empathetic understanding) was couched in terms excluding *Erklärung* (explanation) and *Wissen* (scientific knowledge). Industry, commerce, urbanization, and economic growth were to be regarded as somehow "inhuman" or at least uninteresting for the humanist. Technology, far from being the emancipatory cry of *homo faber,* was construed as evil incarnate. Humanist scholars sought truth in the direct sensory experience of nature and landscape; many, too, directed attention to the intricacies of classical drama and history.

From the late eighteenth century on, however, the most exciting fresh perspectives on the earth as home for mankind were offered by scholars who ignored those Faustian boundaries set down by the *magna carta* for the humanities. Natural historians, botanists, explorers, and poets refused to study humanity in ways that excluded natural science or divinities (Linnaeus 1734; Hutton 1795). Burnet's *Telluris Theoria Sacra* (1680–90) catalyzed further insight into geological time (G. H. Davies 1968; S. J. Gould 1987), as

John Ray's reflections on the hydrological cycle and the wisdom of God (Tuan 1968) eventually paved the way for revolutionary theories about the circulation of water through the body of earth and the circulation of blood in the human body (Mills 1982). The Victorian era heralded a wave of cultural self-confidence and superiority and an emphasis on proper humanist education for citizens of Empire. It witnessed passion over the issues of humanitarianism and noblesse oblige. On the forward march of science and technology, however, many humanists took a cautious, if not reactionary, stance.

Geography, now also claiming academic status within the universities and schools of burgeoning states and would-be empires, identified itself as a science: not only a science equipped to deliver *Erdkunde* but also an exciting practical art. Captivating indeed were images of the geographer as explorer of polar regions north and south, of the Asian heartland, the African jungle, or the North American prairie and desert; and as surveyor, accountant, or reporter on resources, livelihoods, cities, and ports. Closer to home, there was the image of the geographer as master teacher capable of integrating insights from a great variety of other disciplines and providing schoolchildren with edifying pictures of the world and its peoples, as well as understanding of their home regions and nations. Yet, on the eve of its establishment as a formal academic field, geography sought to be counted among the sciences rather than the humanities. The issue for the founders was to carve out a legitimate domain of inquiry on the differentiation of the earth's surface, and to include humanity as one of the formative elements in this drama. How might it confront or circumvent the two powerful orthodoxies of the day, namely, geology and neo-Darwinian natural science, on the one hand, and theology or theological versions of history and nature, on the other? Geography's acceptance or rejection as a university department perhaps had less to do with the logic of its inquiry methods than with the *Zeitgeisten* of its sponsors and audiences.

Despite overt proclamations about being a science, early twentieth-century geography witnessed many creative encounters with the humanities. Of these, the Vidalian school of *la géographie humaine* serves as an exemplar, successfully appealing to contemporary political and pedagogical interests (Claval 1964; Buttimer 1971; Berdoulay 1981). "Geography is a science of places not of men," the oft-cited proclamation of Vidal ran, but the discipline retained a solid anchoring in history as well as in geology. French geography introduced a fresh approach to regional life, with artistically woven accounts that revealed the dynamic interplay of civilization and biosphere. Keenly attuned to the temporal and ecological perspectives pioneered by von Humboldt, Ritter, Ratzel, and others, this school avoided environmental determinism, emphasizing the creative ingenuity of people in developing their *genres de vie* in various regions. To the determinism of natural law, valuable as this was in elucidating the dynamics of relief, climate, and hydrology, they juxtaposed the contingencies of history. By history, one did not imply the officially documented histories (the great tradition) of kings, princes, or elites; folk history (the little tradition), already legitimized by Herder and Michelet, was far closer to the geographical spirit.

La géographie humaine inspired numerous lines of further inquiry into, for example, the sense of place, regional personality, landscape morphogenesis, and bioecology—themes often applauded more enthusiastically abroad than at home (Hardy 1939; Dardel 1952). The success of its thought styles and practice may indeed be due to their fit with the realities of early twentieth-century French rural life and the perceived realities of *France d'Outre Mer*. Its effective counterarguments against critics of the Durkheimian school meant that doors to sociology would remain tightly shut until midcentury (Febvre 1922; see also Berdoulay 1981). The value placed on synthesis and the "art of description" predicated a less than rigorous approach to analysis. Tensions between ideational and artifactual interpretations of landscape, between ideographic and nomothetic, between scholarly and applied, allowed for some intellectual dynamism, but ultimately there was little philosophical reflection or self-critique. *La géographie humaine* provided a model that fitted the realities of early twentieth-century France so well that, in the opinion of many, it became an orthodoxy that was difficult to transcend once those empirical realities had changed.

Many other doors between geography and the humanities were also opened, even during the early period of discipline making, and have continued to admit mutually creative encounters throughout the twentieth century. For example, Banse's ideas on regional spirit, J. G. Granö's *Reine Geographie,* and the *Heimatkunde* of Nordic and Baltic countries all illustrated the values of maintaining a strong humanist spirit in geographical description (J. G. Granö 1929; E. Kant 1934). In North America, the writings of Nathaniel Shaler and John Kirtland Wright stand out as pioneering examples (Koelsch 1979; Livingstone 1982). The record of the Berkeley School has been acclaimed; and its founder, Carl Sauer, also consistently proclaimed that geography was an earth science, *not* a social science (Leighly in PG, 80–89). Before addressing the issue of humanism in the twentieth century, however, there is still another strand to be picked up in the historical record, namely, the concern about the human condition.

Humanitarianism (*Ergon*)

Human creativity in the Western tradition has often been associated with the desire to solve problems and to improve the human condition. It is in the context of *ergon* that ideological conflict has been most dramatic. Leonardo da Vinci's discoveries were explicitly made in the service of princes and potentates; breakthrough inventions in cartography and navigation have been possible and plausible because they served the expansionary ambitions of nations and commercial magnates. Perennially admired works of art and architecture were born in the aura of *laudatio* for the egotistical vanity of ruling classes; but masterpieces of literature, engineering, and music have also sought to stimulate awareness of human poverty and social injustice.

Ergon in geography has traditionally faced an enduring antagonism of stances on the human condition, between scientific humanism and humanitarianism. This tension may eventually be traceable to Socratic times and certainly to the contrasts between Stoic and Epicurean views of creation. It was to the emancipatory dreams of humanity that eighteenth-century scientific humanists addressed their efforts. Human reason, exercised with Cartesian discipline, would deliver humanity from the snares of superstition, myopia, and dogmatism. While humanist academies sent Jesuit padres to launch pioneering efforts of applied geography in the Guarani *reducciones* in the mid–1700s, D'Alembert and others created the *Encyclopédie* to affirm the superiority of human reason over ecclesiastical authority. Auguste Comte was the self-appointed high priest of the *Eglise de l'Humanité* (Comte 1830–42). Scientists and engineers, from the eighteenth through the twentieth centuries, have envisioned rational transformations of landscapes, circulation routes, housing, and industry, convinced that they were contributing to the progress of humanity.

From the humanities, however, another note has been sounded. Oppression and misery following the Industrial Revolution in Europe and America called forth a concern about actual life conditions and diverse forms of humanitarianism. A growing sensitivity to issues of social injustice and inequality was also emerging from disciples of the Hegelian tradition. From lands more directly affected by rapid economic and political transformations came sensitive accounts, in fiction and landscape art, of environmental experience and the daily living conditions of regions and places throughout the nineteenth century. Victorian England, postrevolutionary France, late czarist Russia, and New England all produced vivid, provocative insights into the lived geography of their times. Novelists such as Balzac, Zola, Dickens, and Harriet Beecher Stowe delivered messages equally as effective in pricking the public conscience as were the more scientifically based theories of social history brilliantly expounded by Engels and Marx.

All affiliates of the humanities did not participate in this concern. Many looked askance at such involvement, considering it well beyond the scope (or beneath the dignity) of the proper humanist scholar. The cold war between scholarship and social activism has more than occasionally heated up in academic circles. Among activists, the ideological impasse also lingered. Liberal practices of humanitarian help, charity, helping-the-poor-to-help-themselves, were dogmatically denounced by the Young Turks of scientific humanism, particularly those of socialist convictions. Disciples of Marx, Comte, and Condorcet argued that energies should be addressed to the reconstruction of society, to the elimination of what they regarded as root causes of poverty and injustice. In this they relied on the tools of science and on the best that social engineering had to offer, so that externally imposed constraints on individuals could be removed. From anarchist as well as conservative, theist as well as atheist, the late nineteenth and early twentieth centuries heard many voices on *la condition humaine* and many claims of the potential contribution of the scholar.

To such appeals the geographer has generally responded in terms of the externalities of life: roads, boundaries, industrial location, or the rationality of settlement structures. Concern about the quality of life, social injustice, poverty, or pathology was regarded as more or less the business of the social worker, sociologist, or preacher. Yet in the writings of nineteenth-century geographers who drew inspiration from Humboldt, such as Kropotkin and Reclus, one finds not only keen concern about daily living conditions but also—and perhaps more importantly—convictions about the creative potential of people to seek solutions to their own problems in cooperative and collegial ways at the grass-roots level (Reclus 1877; Kropotkin 1898, 1902; Breitbart 1981; Dunbar 1981). The exhaustive field surveys of working-class families in rural France by Frédéric Le Play and his associates no doubt lent inspiration to the humane concerns of Jean Brunhes and Pierre Deffontaines virtually a century later, as they would to British and American rural sociologists in the twentieth century.

With deep concern over the human condition, nineteenth-century geographers launched an impressive research and pedagogical program during Spain's period of *regeneracionismo* (Gomez-Mendoza and Cantero 1992). Through a creative flight of metaphorical ingenuity, geography flashed forth as Phoenix in a nation debilitated by the loss of foreign possessions and by the mismanagement of resources at home, particularly water. An integrated plan for geographic *paideia* and *ergon* envisioned that the reclamation and rationalization of irrigation systems could lead to a regeneration of Spanish soil, while local education and the improvement of life conditions in particular regions could lead to the revitalization of the Spanish soul. Hypotheses to be explored about essential differences in the claims of humanism in Latin and Anglo-Saxon lands, in Reformed and non-Reformed Christian worlds, and in capitalist and socialist regimes across the face of Europe are tempting indeed (Merleau-Ponty 1947; Teilhard de Chardin 1955, 1959; Lebret 1961; Corbridge 1990).

In North America, the nineteenth and early twentieth centuries enjoyed a very different kind of intellectual and social climate. Pragmatists espoused a common-sense approach to humanism. Agnostic and hostile toward a priori theories about human nature, they simply assumed that people everywhere participated in a common humanity (see Chapter 6). "Man as Measure," the old slogan of Protagoras and Pope, now implied a radically empiricist attitude to thought and life. Whitman and Thoreau had already sought to unburden humanity from the cultural encrustations of history in the nineteenth century; like their romantic forebears, they pleaded for a return to the simplicity of nature itself. Like European existentialists later, American pragmatists promoted an attitude of openness to lived experience and intellectual curiosity about the concrete realities of everyday life. This, it was felt, could help make knowledge more useful to society (James [1907] 1955; Dewey 1925; S. Smith 1984).

The Twentieth-Century Contextual Turn

An explicit recognition of *context,* of environmental and societal circumstances in shaping thought, language, and action, as well as individual life journeys, can be identified as one of the major motifs in the scholarly world during the latter years of the twentieth century (see Chapter 6). It has confronted each of the four faces of humanism touched on in this chapter with a major paradox. With respect to *humanitas,* the exaggerated claims of Promethean individualism evoked stormy protest and eventually proclamations of the death of the human subject. In the venerable debates over subjectivity and objectivity, reason and rationality, sociological rather than epistemological insights were evoked. Contextual light was eventually cast on that peculiar ambivalence detectable among academics toward knowledge and life: passionate declarations about the values of intellectual freedom, balanced against an equally passionate commitment to scientific theories designed to prove how determined everything was. Traditional boundaries separating the sciences and the humanities were also eroded and new alliances emerged, reflecting radical reorientations as well as changing priorities of research funding. Traditional notions about humanitarianism and scientific humanism were also attacked: world wars, the atrocities of so-called civilized peoples, environmental destruction, and global terrorism were hard to reconcile with traditional Western assumptions about *humanitas*.

"Humanism places its faith in humankind," the argument ran, "so that for the continuing worsening human misery . . . it has no such satisfactory explanation, only excuses, lies, evasions, and utopian promises" (Ehrenfeld 1978, 229). The attack comes not only because of reason's apparent inability to solve problems but also because of the tendency to identify reason with scientific rationality (Relph 1981). From the humanitarian viewpoint, too, there has been much disillusionment. Contradictions between ethos and structure in many missionary endeavors and aid programs of voluntary organizations have stirred the political will to handle social problems in a seemingly more positive way via public rather than private sponsorship. Voluntary action became morally suspect in settings where state-run welfare and union definitions of work spread their Faustian mantle over the human condition.

In the 1940s, Heidegger noted that the *humanitas* of *humanus* had been determined from the view of an already established interpretation of nature and of history ever since Roman times. Western humanism, he argued, was thus frozen in its own metaphysical stance. "Humanism does not ask. . . . for the relation of Being to the essence of man, it even impedes this question" (1947, 211). He pleaded for a return to those broader issues of being and becoming which were neglected in a knowledge enterprise steered by such anthropocentric biases (Heidegger 1954). He called for a sensitive, caring, patient listening to reality—letting reality reveal itself in its own terms—rather than seeking to grasp reality in the language of preconceived models. Heidegger's critique of Western humanism satirized the inherited separation of epistemology and ontology (of knowledge and being), of the intellectual and moral dimensions of thought and action.

As Nietzsche, Marx, and Freud in the late nineteenth and early twentieth centuries, so Husserl, Heidegger, and Habermas in the twentieth have voiced radical critiques of the Enlightenment legacy. Of the many issues raised in their work, those that touch most closely on the central aims of this book include the fragmentation of knowledge and the question of foundations, the human subject as author and tensions of agency versus structure, and the issue of language and communication.

The Fragmentation of Knowledge

A keenly felt dilemma in the latter half of the twentieth century, and perhaps the most dramatic legacy of the Enlightenment's Faust, has been that of functional specialization and ultimately fragmentation of knowledge. Approaches to this challenge reveal some of the ideological impasses between humanist and antihumanist. The Frankfurt School, still clinging to a faith in Enlightenment values, articulated its brilliant critique of "instrumental reason" (Horkheimer and Adorno 1947), exploring integrated approaches to the theoretical, technical, and practical aspects of knowledge and emphasizing the cardinal function of communication (Marcuse 1972; Habermas 1976). Others, more Kantian in their diagnoses, sought epistemological foundations for a more integral understanding of reality (Cassirer 1955; Schrag 1980). Several practical steps were also attempted. Programs of "humanistic studies" were initiated in many American universities during the 1970s with the explicit aim of fostering cooperation among researchers in diverse fields and also of rendering the humanities "relevant" for elucidating problems in modern society. A lively exchange did occur between humanists and social scientists, for indeed many scholars—psychologists, anthropologists, geographers, and historians—had already become disaffected from (positivist) scientism in the sixties. By the seventies humanist scholars were showing a greater interest in the societal or contextual aspects of literary and artistic works. Thomas Kuhn's theory of scientific revolutions offered alternative perspectives on intellectual history, and literary critics showed a growing curiosity about context in the unfolding of the classical *chef d'oeuvres* (Kuhn 1970). The correct interpretation of texts was no longer regarded as a matter for the author or any individual reader, but rather for a community of readers tuning themselves into the larger "conversation of mankind" (Fish 1981). Classic texts should be read as "people's attempts to solve problems, to work out the potentialities of the languages and activities available to them . . . by transcending the vocabulary in which these problems were posed" (Rorty 1982, 9). Humanists and scientists could meet in their common concern about the human condition (Frye 1981). Knowledge was to be regarded as a social artifact rather than as a mental construct (Bourdieu 1977; Mendelsohn 1977). Disciplinary practices should be understood ethnographically, as ways of life rather than simply ways of thinking (Geertz 1983, 147–66).

Human Agency versus Structure

On the protracted debates over humanist versus scientific modes of knowing, the ideological tension between freedom and determinism played itself out in the battle royal between positivists and existentialists. During the 1960s, however, some of the basic assumptions of both camps came under attack. The issue for Lévi-Strauss was "not to constitute but to dissolve man," to eliminate "particular, finite, historical subjectivity" (1966, 365). Mind itself would be spared—disembodied mind, no doubt, "human mind, unconcerned with the identity of its occasional bearers." During the heady sixties, Althusser preached, "We can only know something about man under the absolute condition that the philosophical (theoretical) myth about man is reduced to ashes" (1965, 2:179). The structuralist wave from the late 1960s on engendered agnostic attitudes toward such classical assumptions as personal authorship, intentionality, and meaning. Other qualities of human nature such as freedom, responsibility, and compassion were also satirized. Platonic foundations for reason yielded to Heraclitean notions of flux and unpredictability. "Plato's conception of Reason and his realistic conception of Objective Truth," Rorty claimed, "are both forms of what Nietzsche called the 'longest lie'—the lie that there is something beyond mankind to which it is our duty to be faithful" (1982, 2). Philosophy should be abandoned and theory of science or literary criticism embraced. Everything, after all, could now be regarded as *écriture*.

LANGUAGE AND COMMUNICATION

Language, as metaphor for reality as a whole, has long been central to the structuralist position. By the mid-twentieth century, traditional notions of correspondence between symbol and reality were ultimately to be undermined (Foucault 1966; Kunze 1984). More than a century ago, Mallarmé had sought to dramatize the noncorrespondence of words and the objects to which words referred with the metaphor *le coup de dés* (roll of the dice).[6] Wittgenstein noted that the language *in* which one made statements *about* language was but an elaborate rule-system which could be changed at any time. Concerns about "foundations" for knowledge were eventually rejected as just one more feature of a tradition that was caught up in its own unexamined presuppositions (Feyerabend 1961; Elzinga 1980). Rorty traced the movement away from epistemology to hermeneutics in twentieth-century approaches to knowledge (1979). Yet even within the hermeneutic movement, the late 1970s witnessed fundamental antinomies with respect to humanism. For some the central puzzle was that of evaluating interpretations and reaching toward mutual understanding (Gadamer 1965; Geertz 1983). For others it seemed more important to decipher the architecture of those very language structures that served as prisons for human discourse. "A voyage of nothingness and to nothing" is how Steiner characterizes late twentieth-century stances on language and reality, "an echo in both science and the arts, in exact theory and in poetics, of the proposition that 'nothing shall come of nothing,' a statement about the final, incomprehensible, but *expressible* mystery of energy in and out of absolute zero" (1987, 22).

The faces of postmodernism are many—some weary, some hopeful, but all in one way or another expressive of Narcissus (Jameson 1983; Kearney 1988; Folch-Serra 1989). Reflections on the two decades since 1968 have revealed the many irreconcilable ambitions of a would-be Phoenix generation, the cooptation of young idealists into national bureaucracies and bourgeois clubs (Hocquenghem 1986), successful marketing of structuralism among the ranks of former idealists, radical position changes by those who set out to guillotine the human subject (Lyotard 1984; Dreyfus and Rabinow 1984; Ferry and Renaut 1985). By the 1970s, sharp lines were being drawn between humanist and antihumanist; those moral and intellectual distinctions that were already evident in the spirit of the 1960s now became oppositions. Several irreconcilable ambitions were built into the events of that decade, an incompatibility of ideals and proposals for practical action.[7]

From the humanist vantage point, the most heartening signs are those of longing for emancipation from the encrustations of inherited forms of dogma and practice (Frye 1981; Kohak 1984; Steiner 1987; Kearney 1988; Midgley 1989). The human subject has returned, and even within the home of structuralism there has been a mellowing of those dogmatic positions of the 1960s (Ferry and Renaut 1985). "Poets sing of the absence," Heidegger once remarked, "for they have been touched by presence." Emerging from deconstruction is a sharper sense of the absence of creativity and hope, of anything that might transcend the frameworks of a Frankensteinian world.[8]

The prospect for the humanities in the late twentieth century thus echoes those classical tensions between Gnostics and Socratics (see Chapter 6), between images of reality as being in perpetual flux versus images of self-aware human subjects seeking rational understanding of the world. Typically, however, such tensions can scarcely lead to a creative outcome until broader horizons for knowledge and life can be perceived by both. The present global crisis, and the daring images projected by Gaia, may well be the catalysts for such an expanded horizon (Myers 1982; Lovelock 1979). Neither the inherited divisions between intellectual and moral realms of discourse nor those academic fences between humanities, divinities, and natural science block creative imagination today (Eco 1986). Education for moral reasoning, Harvey Cox claimed, could be a central function for the humanities to "nourish human beings capable of passionate imagination, rigorous reflection, reasoned choice and moral courage" (1985). Is this the clarion call of Clairvaux (March and Overwold 1985)?

From Renaissance times on, Western humanism adopted a Janus-like stance on freedom and creativity, which was to mark the full adventure of modern times (Kearney 1988). Intellectual faculties stand out as the perennially prized quality of humanness. Occidental beliefs—Socratic, Judeo-Christian, Latin, and Anglo-Saxon—bent to accommodate and often applaud the Promethean myth of conquest: mankind as superior to other life forms on the earth, with a mission to dominate or shepherd it; mankind as maker of tools and technology to overcome the barriers of distance or disease; mankind as responsible for choices between good and evil. The refrains of *Antigone* persist in Western intellectual history.

As the traditional anthropocentrism of Western humanism came under attack from various corners of the academic world, and environmental issues appeared in public press and debate, the famous Baconian slogan, *natura nisi parendo vincitur* (one conquers nature only by obeying her [Bacon 1620]), took on a new complexion in the 1970s. "Human nature," the "forgotten paradigm," according to Morin, cannot be understood without understanding the complex interplay of culture and nature through history (1973). Exposing the inadequacies of a knowledge enterprise founded on specialized fields of natural and human sciences, and the various isms generated thereby (biologism, anthropologism, psychologism), he sought to unmask the persistent counterplay of rationality and irrationality in human history. Pelt, with more explicitly environmental concerns, claimed that humans would never discover their own nature until they rediscovered the essential reciprocity of sociality and ecology (1977). And Réné Passet, focusing more directly on the tensions between economics and biology, noted the cardinal differences between mechanist and organicist principles of organization in all elements of the biosphere, including humanity (1979). Common among such recent authors is the desire to evoke awareness of what is lacking in conventional stances on life and knowledge, alongside a plea for more holistic, open perspectives. Reflecting on a century's efforts to elucidate the contextual turning of science and humanities, Georg Henrik von Wright could still define humanism as a basic stance on life, one that might again restore integrity and wholeness to the understanding of humankind, nature, and history (1978).

It is in this late twentieth-century reflective moment that geographers too have begun to evaluate critically their own traditions of thought and practice. And of all the challenges afforded in Western humanism, the Socratic "know thyself" remains central. The journey toward self- and mutual understanding, and the search for identity and integrity in a highly structured world, are both part of the emancipatory quest in our day. Western humanity needs to recuperate its role and identity and to discover more appropriate ways of cooperation with fellow humans in healing a wounded planet. Today one is acutely aware of those ways in which academic disciplines have tended to don the ideological garb of their sponsors and audiences, and of how their internal lifeways mirror changing societal trends within those nation-states that summoned them into existence. As Western geographers increasingly recognize the challenge of opening up a respectful dialogue with colleagues in other cultures, and have finally demonstrated commitment to a global research endeavor on mankind's terrestrial home, some critical reflection on our own traditions is surely in order.

Part 2 of this book addresses this challenge in its exploration of four enduring images of world reality, each negotiating its claims to truth at different moments of Western history, each resonating to the cyclical strains of Phoenix, Faust, and Narcissus. The choice of four images by no means exhausts the full variety of cognitive styles which have been expressed in Western traditions, but each offers illustration of how meaning, metaphor, and milieu have intertwined.

Part II

Four World-views in Western Geography

Introduction

The Way of Metaphor

"LANGUAGE is fossil poetry," Emerson once remarked. "Under the microscope of the etymologist," Max Müller claimed, "almost every word discloses traces of its first metaphorical conception" (Müller 1871). Barfield suggested that "every modern language, with its thousands of abstract terms and its nuances of meaning and associations, is apparently nothing, from beginning to end, but an unconscionable tissue of dead, or petrified metaphors" (Barfield 1952). Landscape, like language, could also be interpreted as sedimentations of different modes of discourse: "Architecture is frozen music," said Goethe, and Vidal de la Blache described the humanized landscape as a "medal struck in the image of a civilization."

Geographical language is thoroughly metaphorical.[1] The sur-"face" of the earth has been described in terms of eyes, nose, mouth, cheek, and profile; it has been named and claimed with terms derived from human anatomy and society. Regions and hamlets have been likened to organisms; roadways and canals, to arteries of circulation. Industrial complexes have been described in terms of mechanical processes steered by growth poles as generators of economic development. Place names, too, whether real or imaginary, such as El Dorado, Mecca, Waterloo, and Eden, symbolize particular kinds of experience (Wright 1966; Jolly 1982). Colonists and new settlers often blithely change place names as a mark of occupance. Geographers have labeled places and territories with the metaphors of science, Cartesian coordinates, and toponomies previously inscribed or "explained" in their home countries. Many geographical terms, such as *drumlin, arrête, kame, cirque,* and *chaparral,* were derived from particular localities and everyday experiences. Once these expressions become sedimented and commonplace in textbooks and atlases, the metaphor is often forgotten and the words assume a literal meaning. A study of metaphor in geography could thus elucidate tensions between the indigenous geographical sense of diverse cultures and groups and the standard lore of geographical science. It could also reveal some of the value assumptions embedded in the genesis of disciplinary thought and practice.

John Kirtland Wright challenged his colleagues to explore "the most fascinating *terrae incognitae* of all, those that lie within the minds and hearts of men" ([1947] 1966, 88). *Geosophy,* he suggested, could unmask varieties of geographical sense among individuals and cultural groups around the world. It could also expose the striking differences between popular conceptions of nature, space, time, and landscape and those formally cultivated within the academic discipline called geography. Awareness of such discrepancies invites a radically different approach to disciplinary history. It has long been recognized, of course, that underlying most scientific theories and the classics of Western art and literature, there are implicit hypotheses about the nature of the world (Barnes 1965; Glacken 1967; Nakamura 1980; Galtung 1981). Geographers have attempted, with varying degrees of success, to document landscape correlates of different world-views. Inquiry into cultural differences in the evaluation and use of natural resources has been invited at various moments in disciplinary history (Sion 1909; Hardy 1939; Wright [1947] 1966; W. Kirk 1951; Lowenthal 1967; Bowden 1980; Saarinen et al. 1984). A generation of enthusiasts has explored "mental maps" and varieties of environmental perceptions (Moore and Golledge 1976; Saarinen et al. 1984). But there is more to the experience of world than the visual and the cartographic. Human beings not only see and cognitively schematize their world views, they feel, believe, hope, love, or hate certain symbols of the world (Tuan 1978; Porteous 1990). The mental map or cognitive schema may not capture the full range of sensory and emotional aspects of people's experience of world. Nor have Western geographers been so successful in promoting better communication or mutual understanding among people who adhere to contrasting world-views.

A treasure of insight can indeed be unlocked via metaphorical rather than literal or rational thinking (Sachs 1979; Black 1962; Tuan 1968, 1978; Livingstone and Harrison 1981; Kunze 1984). Metaphor touches a deeper level of understanding than "paradigm," "model," or "theory" (Ricoeur 1971, 1975). It points to the very process of learning and discovery, to those analogical leaps from the familiar to the unfamiliar which rally imagination and emotion as well as intellect (Langer 1957). Metaphor performs a poetic as well as a conservative function in ordinary language, preserving as well as creating knowledge about actual and potential connections between different realms of reality (Ricoeur 1975; Hofstadter 1979). Variously defined as "the dreamwork of language" (Davidson 1978), the "intellectual link between language and myth" (Cassirer 1946; Malinowski 1955; G. S. Kirk 1970), as literary trope (Jakobson and Halle 1956), as "mode of argument" (H. White 1973), metaphor has aroused curiosity in art, philosophy, music, and history, as well as in the social sciences (Shibles 1971; Sachs 1979; Morgan 1980; Mills 1982). Scientists and philosophers of science tend to focus primarily on the cognitive import of metaphor, on its role as bearer or translator of intellectual meaning (Black 1962; Hesse 1966; Leatherdale 1974; Ortony 1979; Lakoff and Johnson 1980). Yet it was Einstein, one of the greatest scientists of the twentieth century, who claimed that the search for "universal laws from which a picture of the world can be obtained cannot come by pure logic . . . such [laws] can only be reached by intuition" (Popper 1976, 32). The human scientist has much to glean from the mythopoeic and heuristic appeal of metaphor. What attracts the geographer is the capacity of metaphor to facilitate better understanding of the human experience of world, and culturally diverse perceptions of nature, space, time, and social life (Jung 1964; Ferguson 1978; Turner 1980).

Humanity everywhere and at all times has demonstrated ingenuity and "metaphorical imagination" in its modes of life and thought (Mills 1982; Kunze 1984). Giambattista Vico's *Scienza nuova* (1744, par. 209) noted that "the earliest men, as children of the human race . . . felt the need to compose for themselves poetic characters, that is, imaginative genera or universals, in order to reduce, as it were, specific types to certain prototypes or idea portraits, each one to the species resembling it" (cited in Kunze 1984, 172).

As Vico, Herder, Cassirer, Langer, and others have argued, perhaps the most essential quality of humanness is the capacity symbolically to transform reality. Varieties of language and music, architecture and art, boundaries and fences, liturgical and diplomatic styles, all reveal the marvelous diversity in human perceptions of nature, space, and time. "The development of civilizations," Doctorow claimed, "is essentially a progression of metaphors" (1977, 231–32). Varieties of metaphorical imagination apparent in the geographers' own career narratives illustrate the connections between knowledge and life experience, evoking ethnographic as well as epistemological reflection on disciplinary thought and practice (see chapter 1). Metaphor offers a promising catalyst for self- and mutual understanding among individual scholars, and eventually among world civilizations.

Steven Pepper's theory of *World Hypotheses* (1942) offers one potential route for identifying basic metaphors in the history of Western geography. Writing before midcentury, he claimed that there were really only four relatively adequate explanations of reality which have stood the test of time and intellectual scrutiny in the Western world: formism, mechanism, organicism, and contextualism. Each of these "hypotheses" could be regarded as the source of many paradigms, models, and analogues that have guided the analytical frontiers of science and the humanities. Each rests its cognitive claims on a particular theory of truth, and each has a "root metaphor," which is ultimately grounded in common-sense experience:

> A man desiring to understand the world looks about for a clue to its comprehension. He pitches upon some area of common sense fact and tries if he cannot understand other areas in terms of this one. This original area becomes then his basic analogy or root metaphor. He describes as best he can this area, or, if you will, discriminates its structures. A list of its structural characteristics becomes his basic concepts of explanation and description. These we call a set of categories. In terms of these categories he proceeds to study all other areas whether uncriticized or previously criticized; he undertakes to interpret all facts in terms of these categories. In the process he may reinterpret some of his categories—they change and develop. Since the basic root metaphor normally arises out of common sense, a great deal of refinement is required if the categories are to prove adequate for a hypothesis of unlimited scope. (Pepper 1942, 91–92)

This approach derives from the pragmatic naturalism of his teacher, John Dewey. One could regard it as a kind of metaphysics which is concerned with the empirical realm of existence. A philosopher of aesthetics and a historian of ideas, Pepper sought to illustrate his theory in the context of art, ethics, and the history of science.[2] He argued vehemently against the attitudes of "utter skeptic" and "utter dogmatist" with respect to truth claims, but also dismissed eclecticism as philosophically confusing. Though he cited works that exemplify the four root metaphors, he emphasized that the object of the exercise was not to stereotype authors; rather, it was to characterize thought styles. Individual authors might move back and forth among root metaphors in the course of a career, so the conventional practice of stereotyping scholars and schools with rubrics of method or "paradigm" would ultimately be myopic.

How, then, might it be possible to establish any stance from which to regard all of these metaphors objectively? The inevitable hermeneutic circle (Schrag 1980; Palmer 1969) was acknowledged: "It is a puzzle like that of a bottle carrying a label of the picture of that bottle, which picture of that bottle is pictured with a label which pictures the picture of that bottle" (85).

Pepper presented his theory as a "makeshift receptacle in which to gather whatever seems cognitively valid in the various world theories we have" (85). He invited readers to bracket or set aside those habits of thought generated during the positivist era and to open the universe of critical thought beyond the technical (methodological) level. Positivism, he noted, seeks truthful statements via a process of "multiplicative corroboration"; for example, statistical manipulations, pointer readings, tests of significance, and other measures of consensus among observers of data. The perspective underlying the world hypotheses approach was one based on "structural corroboration" (fig. 14). All phenomena perceived by the senses were re-

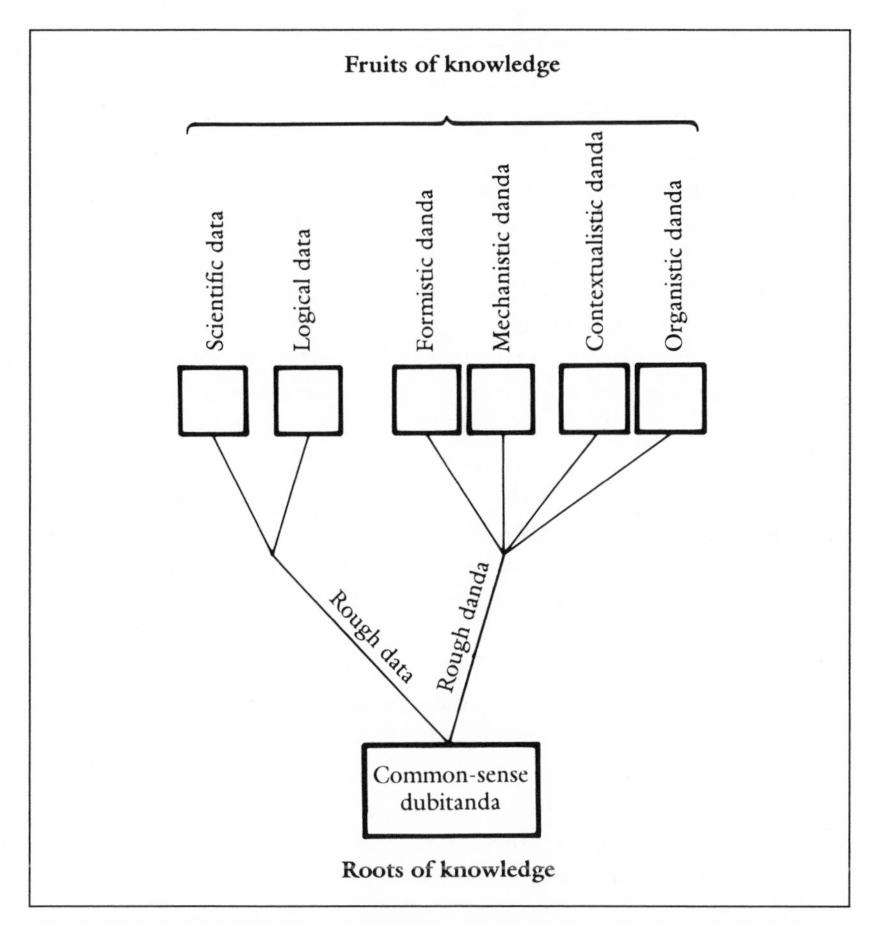

Fig. 14. Roots of knowledge according to the theory of world hypotheses. (After Pepper 1942, 68. Reproduced by permission of the University of California Press.)

garded as *dubitanda* (matters to be doubted). When scrutinized via the structures of potential explanation, they could become *danda* (facts to be given). Whereas the derivation and interpretation of *data* in positivist procedures followed fixed rules, and the credibility of results depended on the consistency of procedure and inference to which all researchers adhered, Pepper's approach suggested that there were four distinct modes of deriving and interpreting *danda*, each ascribing to a distinct theory of truth and generating a set of categories through which its discourse on reality was to be channeled. It would therefore be invalid to subject the refined products (*danda*) of any one to scrutiny via the categories of another.

Formism grounds itself in the common-sense experience of similarity;

its cognitive claims rest on a correspondence theory of truth (Pepper 1942, 151–85). Its world picture is a dispersed one: each form and pattern may be explained in terms of its own nature and appearance. Mechanism takes the common-sense experience of the lever or pump, and its claims to cognitive validity rest on a causal adjustment theory of truth (221–31). Its world picture is an integrated one and, like formism, it is analytically oriented. Organism also offers an integrated picture of the world, its aims being synthetic rather than analytical. It sees all events in the world as more or less concealed process, all eventually reaching maturation in the organic whole. Its cognitive claims rest on a coherence theory of truth (280–314). Finally, contextualism draws its inspiration from the common-sense experience of unique events. It seeks to unravel the textures and strands of processes operative in, or associated with, particular events. Its world-view is also a dispersed one, although its descriptive style is synthetic. It espouses an operational theory of truth (268–79).

Though each of these world theories is autonomous and distinct, there are ways in which they resemble one another, and there are bases for comparing them in terms of adequacy and scope. The universe for both mechanists and organicists is one of total determinateness: a giant machine or an organic whole. As integrative theories, these metaphors do not easily tolerate notions of chance and indeterminateness. In this they are threatened with lack of scope. For the formist and contextualist, on the other hand, the universe is a multitude of facts, loosely scattered and not necessarily determinate of one another: chance, unpredictability, and an indeterminate cosmos are consistent with these two theories. While they are unlimited in scope, they are threatened with lack of adequacy. Formism and mechanism, both analytical theories, tend to complement one another and, in Pepper's words, they "fly to each other's arms for support like animism and mysticism" (146). Contextualism and mechanism are so close that they also tend to complement each other: mechanism gives substance to contextualist analysis, and contextualism gives life and reality to mechanistic syntheses. Contextualism and organicism share a preference for synthetic rather than analytical approaches. They are often so close that they could be seen as part of one theory: one is dispersive, the other integrative.

Like all metaschemes, the Pepper four-part schema has its strengths and limitations. It has held considerable appeal in fields such as psychology, medicine, education, and aesthetics as scholars seek to transcend their disciplinary boundaries in the quest for more integrated understanding of issues (Efron and Herold 1980). The schema has also offered fresh perspectives on the history of science. Hayden White's study of four nineteenth-century master historians (Ranke, Michelet, de Tocqueville, and Burck-

SUMMARY SKETCH OF PEPPER'S WORLD HYPOTHESES (1942)

WORLD HYPOTHESIS	TYPES		CATEGORIES	THEORIES OF TRUTH	ROOT METAPHORS	AUTHORS*
Organicism		IDEAL	1. Fragments of experience 2. Nexuses (connections) 3. Contradictions (gaps, oppositions, actions) 4. Organic whole (resolutions)	Coherence	Organic whole	Schelling Hegel Green Royce
		PROGRESSIVE	which (4) 5. Is implicit in the fragments 6. Transcends previous contradictions 7. Economizes, saves, preserves original fragments			
Formism	Immanent		1. Characters 2. Particulars 3. Participation	Correspondence	Similarity of form	Plato Aristotle Scholastics "Modern Cambridge Realists"
	Transcendent		1. Norms 2. Matter 3. Principles of exemplification			
Mechanism	Discrete	PRIMARY	1. Field of locations 2. Primary qualities (weight, pressure, parts) 3. Laws holding among configurations of primary qualities in the field	Causal adjustment	Lever/machine	Lucretius Galileo Descartes Hobbes Locke Berkeley Hume Reichenbach
	Consolidated	SECONDARY	4. Secondary qualities (Color, texture, smell, etc. which do not affect the functioning of the machine) 5. Principles whereby (4) are connected with (1), (2), and (3) 6. Secondary laws governing relationships among (4)			
Contextualism			A. Quality: (1) Spread of an event (2) Change (3) Degrees of fusion B Texture: (1) Strands of a texture (2) Its context (3) References: (a) linear (b) convergent (c) blocked (d) instrumental	Operational	The historic event	Protagoras Peirce James Bergson Dewey Mead

*Pepper stresses the fact that few authors, with the exception of "dogmatists," have unequivocally exemplified these root metaphors; most creative scholars, he suggested, work their way through several.

Fig. 15. World hypotheses, root metaphors, cognitive claims. (Based on Pepper 1942.)

hardt) used Pepper's idea of world hypotheses to shed light on the implicit precritical grounds on which these authors built their theoretical constructs (1973).[3] The four root metaphors were used to characterize "mode of argument," and correspondences between these and other features of the texts scrutinized—for example, ideological implications and prose style—were explored. White's juxtaposition of schemata, designed to clarify variations in the practice of history, invites parallel studies in the history of Western geography. Evidence from our dialogue project, however, suggests that the overall approach is more appropriate for a characterization of general trends in disciplinary practice than it is for the discernment of individual cognitive styles. As Torsten Hägerstrand cautioned: "I had my Physical Geography without ever hearing about a mechanistic world-view and my History without a word about pragmatism or even Hegelian dialectics. The result is, I believe, that most scholars, working in the separate disciplines remote from philosophy itself, live with a patchwork of concepts and views which do not neatly fit into one clear system from beginning to end. . . . The basic culture is responsible for much of the coherence there is to be found" (personal communication).

Historians of geographic thought might also quaver about Pepper's apparently blithe dismissal of other potential world theories, about his emphasis on the "cognitive juices" of a metaphor, and his virtual concentration on Western thought alone. But his schema of world hypotheses does offer a useful narrative frame within which the story of Western geography could be told. Part 2 of this book sketches these four world-views and their essays at negotiating truth claims in different circumstances—how they reflected, and became reflected in, the material and cultural contexts of successive periods: the world as *mosaic of forms* (Chapter 3), as *mechanical system* (Chapter 4), as *organic whole* (Chapter 5), or as *arena of spontaneous events* (Chapter 6). Through the centuries of recorded history, each of these metaphors in turn sought to deliver its explanations of reality via graphic, mathematical, or literary modes of expression. And the stage on which this drama has unfolded is none other than the variegated surface of the earth. Playwrights in their plots reflect, implicitly or explicitly, the vicissitudes of sponsorship and audience. With the advent of *discipline,* scarcely more than a century ago, a radically new element was introduced into the geographical interpretation of earth reality. Nation-states assumed responsibility not only for playwriting and plot but also for the material contexts and relative status of research, teaching, critical thought, and applied work. Each national tradition would henceforth have its own literal and archivally documentable story to tell about the social construction, language games, and career trajectories of its heroes. But across linguistic and cultural boundaries within the Western world, as the following chapters seek to show, basic varieties of cognitive style could be observed, all of which bear a family resemblance to, or appear to have been derived from, these four root metaphors.

The trilogy of symbolic figures, *Phoenix, Faust, Narcissus,* remains as leitmotif in these accounts. From the moment when geography became established as an academic discipline, however, the other trilogy of themes, *meaning, metaphor, milieu,* offers a more incisive interpretive frame. Together, the two trilogies elucidate the fundamental issue of integration versus integrity in the practice of the discipline.

The story reveals a kaleidoscope of symbolic, aesthetic, and ideological features of Western geography, its anchoring in successive milieus, all reiterating the more fundamental issue of contrasting and irreconcilable worldviews. But metaphor can point beyond itself, directing imaginations toward those mythical and cosmological sources from which conventional explanations of reality spring. Instead of uttering tiresome *mea culpa*s over the Midas touch of applied science and technology, the Western world needs to explore new metaphors, to revive or rediscover other values to guide its relationship to earth and world (Schwenk 1976; Kiliani 1983; Worster 1986). Could not the encounter with *terrae incognitae* be an occasion for such discovery? Could it not also be an occasion for emancipation from those technical interpretations of thought which have characterized much of the Western journey thus far (Heidegger 1971; Gadamer [1965] 1975)?

The root metaphor approach affirms a plurality of potential stances on the diversity of geographical experience, affording the opportunity to observe simultaneously the social, epistemological, and material contexts in which certain types of consensus have been reached. With such an understanding of the past, geographers might be in a better position to encounter and understand the root metaphors of other civilizations.

Chapter Three

World as Mosaic of Forms

> Beat of the traffic, pulse of the phone, the long cycles of the angle of the sun in the sky. Patterns, rhythms. We live by patterns. Intervals. Repetitions. . . . To perceive a pattern means that we have already formed an idea what's next. Rhythms in space. A great scientist said that there's no science without measurement and quantity—but he was wrong, for in science as in life, patterns come even before numbers. . . . The spiral of a snail's shell, the spiral of the great nebula in Andromeda. . . . Folding of rocks and meanders of rivers. Music is *all* pattern—too regular and the music is banal, but a great composer teases our sense of pattern, upsets expectations but then resolves the complexity by reimposing the pattern at a more encompassing level. Symmetry. Broken symmetry. Mathematics itself, in large part, is the recognition and pursuit of patterns in numbers.
>
> —Judson 1980, 28

THE STIRRINGS of geographical awareness among humans can be traced to the recognition of forms and patterns, similarities and differences, among people, places, and events. The origins of language, artistic expression, and philosophical inquiry about the nature of reality can all be traced to curiosity about the variety of tangible forms on the earth. Early accounts of distant lands and journeys took the form of sagas and epic poems, which aimed to reaffirm moral truths as much as to excite intellectual exploration. The *Gilgamesh,* anticipating Homer by over a millennium, created images of cedar forests, mountains, rivers, and monstrous warriors in a tale about good and evil, love and hate, creation and death, earth and heaven. Still today, Western science and philosophy anchor much hope on the analysis and description of form as catalyst for discovery, from global ozone to the human brain (Sheldrake 1981; Judson 1980; Mandelbrot 1983).

Throughout the centuries of recorded human history, civilizations have varied in their interpretations of patterns and forms displayed in nature, landscape, and human life. Megalithic monuments reveal precise knowledge of the phases of the moon. Babylonians considered stars as the "writing of the sky"; Arabs calculated time according to the stars. Varuna, in the Hindu scriptures, shot out the days like arrows; in Scandinavian lore, the *nornorna* (Norns) wove and spun the threads of time, one of them clipping off the measured allotment for each individual. In the Book of Psalms, natural forms were read as manifestations of Yahweh's infinite wisdom; in Buddhist and Hindu teaching, each natural form was regarded as a participant in the universal process of ongoing creation. India, perhaps most especially, has bequeathed elaborate symbolic expressions of the cosmos as a whole, its nature and purpose, as well as the place of each individual within it.

"There are many thousands of worlds comparable to the *sūtras* within a single spade of dust," the Dōgēn (Japan) proclaimed; "within a single [speck of] dust there are innumerable Buddhas. A single stalk of grass and a single tree are both the mind and body [of us and Buddhas]" (Nakamura 1980, 283). Zen gardens and miniaturized natural forms proclaimed the balance and beauty of Tao, or The Way (Spiegelberg 1961, 19). The Meghaduta of Kalidasa ascribed conscious individuality, a real personal life, to all forms of nature; in fact, the poet described aspects of nature that correspond to various human emotions (Kale n.d.; Murray and Murray 1980). Humans everywhere have superimposed their own patterns and rhythms on the world, but nature itself with all its myriad forms of life has continued to transform the patterns around us. "There's more syncopation for the eye and mind in a square mile of Alpine meadow or Sumatran rain forest than in all of Mars and Jupiter with our moon thrown in" (Judson 1980, 29).

Center and Horizon

The classical world of the Mediterranean, meeting ground of Asian, North African, and European civilizations, has bequeathed an enduring legacy in the interpretation of form and pattern. Here indeed, it was believed, lay the center of the world. For some it was the Dome of the Rock in Jerusalem; for others, Delphi; for still others, Alexandria or Byzantium; and to the present day, for many, it is Rome, the Eternal City. Throughout the landscapes of classical Greece, temples, monuments, arenas, and arches witness a highly sophisticated approach to spatial harmony and geometric proportion (fig. 16).

In the forested lands of northern Europe, by way of contrast, Scandinavian myth symbolized the center of the world as a tree, the Yggdrasil, whose roots descended into the tripartite underworld, whose trunk traversed the world of mortals, Midgård, and whose branches ascended into the world of gods, Asgård (fig. 17).

The Upanishads celebrate the eternal Asvatta (fig tree) and the banyan tree, whose roots grow down from its branches and thicken into "pillar-

Fig. 16. Theater at Epidaurus. This theater, dedicated to Dionysus on the wooded slopes of Mount Kynortion, expresses at once the perfection of acoustics, mystical mathematics, and ecologically attuned aesthetics. The built-in harmonies of form were intended to heal as well as to elevate the spirit via its dramatic interpretation of the meaning of life. (Sketch by Bertram Broberg.)

roots," or subsidiary trunks (Cook 1974, plate 64). The North American Sioux built their entire cosmology around the sacred cottonwood tree (Neihardt 1961). Symbols of center are only one part of the story, for rest and movement, like home and horizon, are reciprocal human experiences. As tree and canoe together constitute the central symbols for Vanuatu islanders (Bonnemaison 1985), the Viking ship is an enduring symbol for Swedish space and time alongside Yggdrasil (fig. 27).

The Greeks also traveled widely and were keen observers of nature. The Argonauts certainly penetrated the remote north, and Homer's *Odyssey* (ca. 800 B.C.) provides remarkable detail on the geography of people and places, wars and migrations, and diurnal patterns of activity and repose attuned to different milieus. Two overriding curiosities permeated the classical works of Greek *historia:* Why the differences between peoples? And why were some parts of the earth habitable and others not?

The origins of Western geography are traceable to these same sources. Hekataios of Miletus (ca. 520 B.C.) wrote *Circuit of the Earth* and sketched a map on which details of voyages, coastlines, cities, and peoples could eventually be plotted. The Ionian map had one central axis stretching from the Pillars of Hercules (Gibraltar) through the island of Rhodes and ex-

Fig. 17. Yggdrasil. The tree is watered from the three wells at the spring of Mimir (Remembrance), where Odin once sacrificed an eye for a draft of its wisdom; the springs of the Norns, Urd (Fate, spinning the thread), Verdandi (Being, winding the thread), and Skuld (Necessity, cutting the thread that measures the life lengths of humans); and the Hvergelmer spring where Niogghr, gigantic serpent, perpetually gnaws at the roots of the tree. The Norns represent three faces of the moon goddess, waxing, fullness, and waning—three concerns in the rhythms of life. They sit, presiding over the "irrational" seeds and latencies of germinating forms: water, chaos, and night. The gods live in Asgård, at the upper branches of the tree; humans live in Midgård, at the middle branches. The serpent is the adversary of the eagle who lives on the upper branches, seat of Odin, where he surveys nine worlds covered by the tree. Various creatures live off the tree; a squirrel runs up and down its trunk; other animals eat its branches, leaves, and shoots. Perennial strife therefore unfolds between animals attacking the tree and Norns watering its roots, all within the context of the struggle between eagle and serpent, symbolizing tensions between solar and lunar influences within the cosmos. At Ragnarök, the great tree shakes, destroying gods and world. Yet, concealed within the trunk are the forms of a man and a woman from whose union a whole new race will emerge. (Sketch by Bertram Broberg.)

tending eastward. Later, Eratosthenes (third century B.C.) defined a North-South axis or "prime meridian" from the mouth of the Dniepr in the north, through Byzantium, to Alexandria, where his famous school was located, and where, a few centuries later Ptolemy, father of Western scientific geography, was born.

Through subsequent efforts by Alexandrian scholars such as Hipparchos and Posidonius, all the rudiments of latitude and longitude were developed, and the seeds of further explorations into map projections were sown. *Geographia,* in contrast to mere *historia,* had already assumed a deductive stance on how the cosmos should be described: it sought a geometric framework within which diverse information could be recorded and analyzed. Thus, previously unassimilated masses of chorographic and topographic descriptions could eventually be assembled on a new grid and examined in terms of meaning and order in the world, the other great concern of Greek scholars.

These geographical curiosities can scarcely be appreciated until they are placed in the context of deeper ontological concerns about the nature of the universe, about the cosmological ordering of things, and possible teleological meanings and purposes in creation (Glacken 1967). In the sixth century B.C., Thales proposed that "all is water"; Anaxagoras (ca. 500–428 B.C.) argued that it was air, or intelligence, that provided the fundamental ordering principle. Anaximander (611–547 B.C.) held that the universe was indeed governed by law but rejected any single-element explanation. It was Empedocles (492–432) who developed an organicist interpretation of the universe based on the four basic elements—fire, earth, air, and water—whose interactions were steered by two others, namely, the unifying power of love and the dividing power of hate.

AIR
HOT AND MOIST
BLOOD

FIRE
HOT AND DRY
BILE

EARTH
COLD AND DRY
MELANCHOLY

WATER
COLD AND MOIST
PHLEGM

Like analogies to the human body, this doctrine of the four elements catalyzed virtually all attempts to describe reality as a whole. For the formist, however, its appeal lay in the hypothesis that each element had its counterpart in the human body's "humors." The Hippocratic School of medicine developed the most elegant application of this theory (see Chapter 6). Health was construed as the harmonious blending and balancing of powers—the proper proportioning of the humors. Since good health was naturally contingent on some aspect of the physical environment—temperature, precipitation, soil quality, or other concurrent conditions—every doctor needed an education in geography and philosophy before initiating studies of medicine. The impact of climate on human health and culture remained an enduring puzzle among Western geographers, and the doctrine of the four elements was an underlying theme of Renaissance formism. For the Greeks, it motivated a mapping of the known world in terms of *klimata,* or zones of the *oecoumene* (humanly inhabited world) within which distinct patterns of culture could be represented and studied. The underlying motif was the principle of correspondence, one that was to lead to remarkable achievements in the understanding of diseases and cures by Arab physicians; to Scholastic treatises on virtue and the formation of personality; and eventually to the quest for the optimal climate for human creativity and the ideal political order and laws for people of different environmental milieus.

Another idea prevalent in classical Greece sources was that the universe was created by a wise and benevolent maker. Herodotus (400 B.C.) noted: "There is a divine providence, with a kind of wisdom to it, as one might guess, according to which whatever is cowardly of spirit and edible should be prolific in progeny, so that, with all the eating of them, they should not fail to exist; while things that are savage and inflict pain are infertile" (Herodotus, bk. 3, 108).

Of the various hypotheses about the meaning and heuristic value of studying form in the classical world, Plato and Aristotle have certainly had the most enduring influence on subsequent thought. In the Socratic tradition, visible forms were interpreted as surface expressions of underlying formative processes—causes, natural laws—discoverable by *logos*. On the nature of such transcendent processes, a clear distinction could be made between Platonic and Aristotelian world-views.

Plato viewed creation as the work of an intelligent, good, reasoning, and divine artisan. The cosmos, it is argued, is obviously in process of becoming, and therefore must have a cause. In the *Timaeus,* two models are posed for the design of the cosmos: one ideal and eternal (existent, and not becoming), the other real and transitory (always becoming, never existent) (Plato, Bury trans., 1952). If the divine artisan used the eternal as model, it would be beautiful; if he used the created one, it would not be beautiful. He therefore chose the beautiful. In the cosmos, the four elements were held together by friendships ordained by God. The Platonic artisan is really *mind* imposing itself on reluctant matter: a theme that recurs again and again throughout Western formism (Augustine 1972, 418–20). For at least the first millennium of Christian Platonism, too, the quest for truth was not separable from the quest for goodness and beauty (figs. 18, 19). Now, for Aristotle, the artisan of creation may have chosen a different model, namely, one in which individual beings continued as co-creators, as it were, each working toward their appropriate ends within nature. The final cause was the rational end implicit in the formative processes of nature itself: "Everything that Nature makes is means to an end. For just as human creations are the products of art, so living objects are manifestly the products of an analogous cause or principle, not external but internal, derived like the hot and the cold from the environing universe" (Aristotle 1941, *Parts of Animals*, bk 1, chap. 1, 12–16). "The end of the species is internal to that species; its end is simply to be that kind of thing, or, more definitely, to grow and reproduce its own kind, to have sensation, and to move, as freely and efficiently as the conditions of its existence—its habitat, for instance—allow" (cited in Glacken 1967, 48).

The fundamental distinction between Platonic and Aristotelian interpretations of form could be traced throughout the millennium that followed. Aristotelian ideas bore fruit in the neo-Persian empire of the Sassanians, in the monasteries of Byzantium, and especially in the Arab world. Platonic ideas tended to dominate within Christendom until the eve of the Renaissance in the twelfth and thirteenth centuries.

A Correspondence Theory of Truth

Already among classical scholars, one can discern the rudiments of what Pepper labels the "world hypothesis" of formism, and its "root metaphor," form, which rests its cognitive claims on a correspondence theory of truth (1942, 151–85). Whether in philosophy, science, or art, the central challenge of formism has been to discriminate similarities and differences, to classify and typify such differences, and then to represent (via prose, numbers, maps, art, or sound) the forms that have been perceived by the senses. As world-view, it has been one of the most enduring in Western intellectual history, and has been certainly the most pervasive in the history of geographical thought.

Pepper identified two kinds of formism: immanent and transcendent. Immanent formism implies a straightforward description, via symbolic representation, of phenomena. It identifies certain (1) characters, (2) particulars, and (3) levels of participation, or the tie between characters and partic-

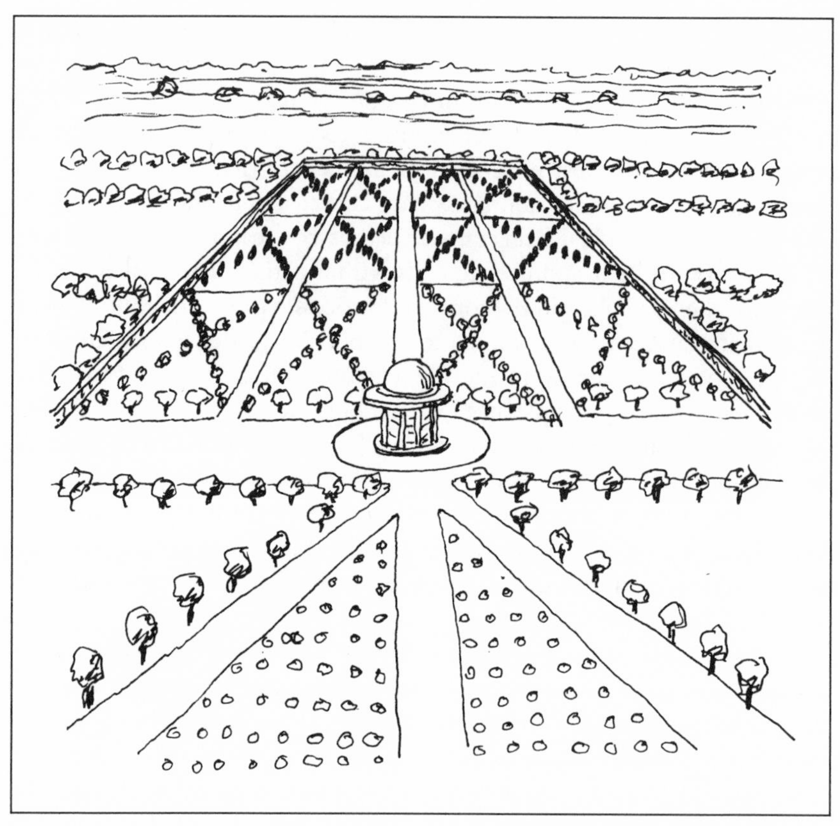

Fig. 18. Artisan analogy. Visible forms imply an underlying plan. As the soul informs the body, so too the divine artisan's plan informs all of creation. (Sketch by Bertram Broberg.)

ulars. Pepper's illustration (Fig. 20) indicates the direct relevance of this line of reasoning for the cartographer.

A "class" is a collection of particulars which participate in one or more characters, and classification systems proceed from the more general to the less general. Classes are expressions of the actual working of the abovementioned three categories in the real world.

The appeal of formism to cartographers down the centuries should be self-evident. But mapmakers from the earliest times faced a twofold task: one was to represent phenomena symbolically, as illustration and contextual grounding for *historia* (à la Herodotus); the other challenge was to devise a mathematically based grid system to serve as a potentially universal framework for the study of the earth or the cosmos as a whole (à la Ptolemy). While the former maintained an inductive approach and aim at contextual

Fig. 19. Natural law. Norms are intrinsic to nature itself, each form in process of becoming, for example, the oak tree. The norm involved here transcends the particular form: a given oak tree may be influenced by its milieu and inheritance; few ever achieve the perfect norm. (Sketch by Bertram Broberg.)

sensitivity to local detail and contextual variety among the phenomena observed, the latter pointed toward abstract models of the earth and a deductive science of location and spatial organization.

Transcendent formism went beyond the level of description and representation. It sought *norms,* or models, yielding explanation of patterns. Three further categories would thus distinguish it from immanent formism: (1) norms, (2) matter for exemplifying those norms, and (3) principles of exemplification which materialize the norms. Within transcendent formism, as Pepper defines it, one could again discern the distinction between Platonic and Aristotelian stances on being and becoming. What distinguishes the immanent and transcendent types of formism could be roughly translated into the distinctions between inductive and deductive types of inquiry

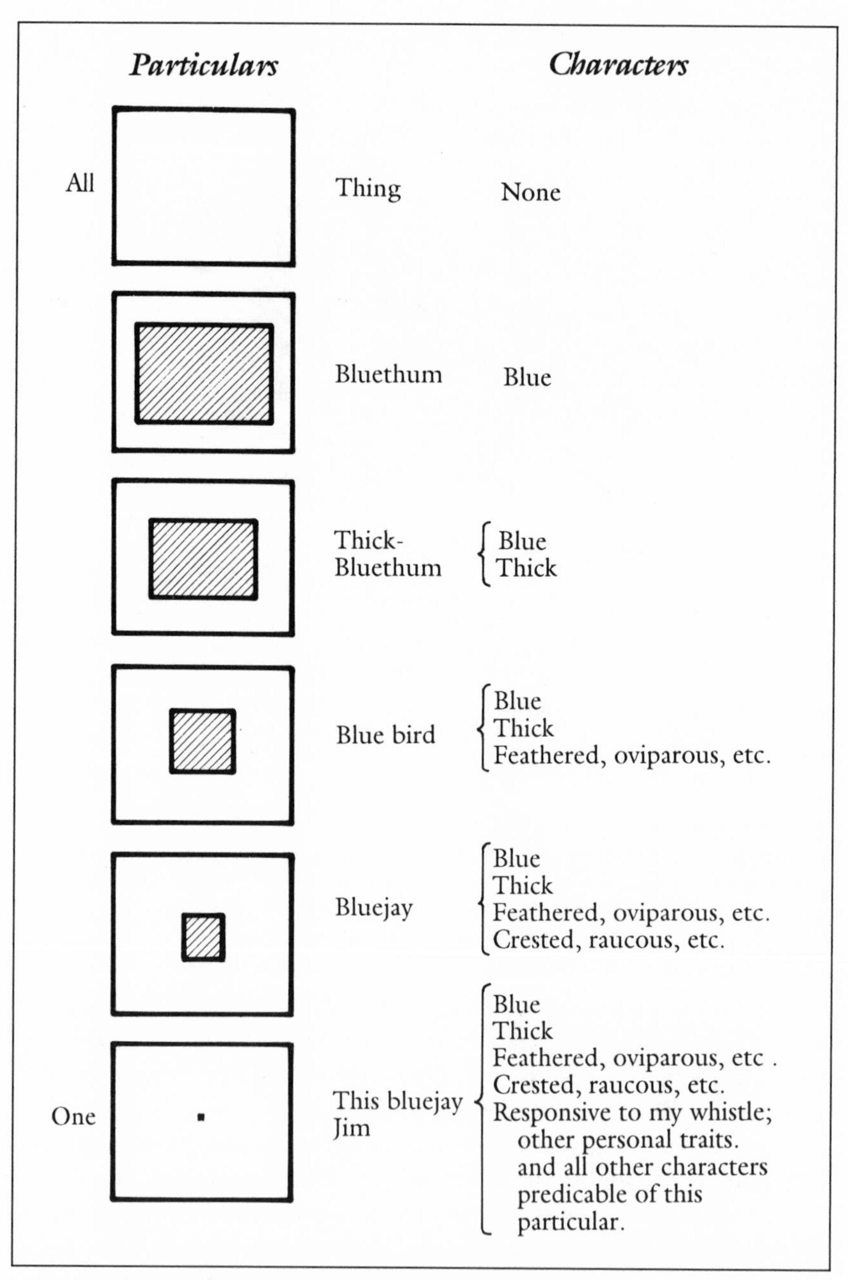

Fig. 20. The logic of classification. (After Pepper 1942, 160. Reproduced by permission of the University of California Press.)

in geography, a distinction that could be observed through all subsequent cartographic and geographic inquiry. Both yielded ample scope for controversy within philosophical and theological circles throughout the medieval period in Europe and North Africa.

Realism versus Nominalism in Medieval Times

The so-called Dark Ages, stretching from the decline of the Roman Empire through the eve of the Renaissance, witnessed many probings into issues of form and pattern. There was a strong current of anti-intellectualism among patristic writers such as Tertullian and even Jerome ("God has spoken, what need have we of further knowledge?"). The Christian cosmographer Cosmas (fifth century) endeavored to demonstrate that the world was flat. There were tensions, too, and occasional hostility between students of Greek philosophy and students of Christian theology. Augustine (A.D. 354–430) stands out as the potential synthesizer of these two streams, and the questions he raised about space, time, memory, being, and nonbeing are still speculated upon in twentieth-century philosophy. Influenced by the neo-Platonist Plotinus and his disciple, Porphyry (third century A.D.), Augustine's style of "transcendent formism" had a decidedly theological flavor. In an oft-cited passage, he regards the forms of nature as a book, a text to be read for the edification of humanity: "Some people, in order to discover God, read books . . . but there is a great book: the very appearance of created things. Look about you. Look below you. Note it, read it. God, whom you want to discover, never wrote that book with ink; instead He set before your eyes the things that He has made. Can you ask for a louder voice than that?" (cited in Glacken 1967, 204).

Underlying all visible forms lay the intentional act of an author, in this case a divine artisan. Throughout the eighteenth century, in the voyages of missionaries and explorers, in the curricula of Jesuit (humanist) academies, in the physical geography of Ritter and Guyot, a similar motif led to the promotion of geographical knowledge. The Aristotelian model, seeking universal norms in the nature of things themselves, would find more enthusiastic expression among Muslim, Byzantine, and Persian circles, eventually rediscovered via Spanish Moors by Albert the Great (1193–1280) and his student, Thomas Aquinas (1225–74), who created a new synthesis of science and philosophy, of faith and reason. For Aquinas, all natural forms, including human reason, displayed qualities of activity and passivity; each had potentiality that could be actualized by the twin processes of rationality and divine grace. Existence (being) and truth were coterminous and inseparable—a principle later to be expounded by Kierkegaard, Heidegger, and twentieth-century existentialists.

The Middle Ages could be regarded, in many ways, as the heyday of formism, and the parallels with the late twentieth century are striking (Eco 1986). Admittedly, it was an age of faith, and puzzles as yet unsolvable via the logic of empirical inquiry could be readily "solved" by reference to holy writ or *Roma locuta est* (Gilson 1950). Yet there were types of formal inquiry of a clearly secular nature that broached questions in ways that still engage intellectual imaginations today. The central puzzle in medieval philosophy was that of *universals,* couched in a succinct and provocative manner by Porphyry in the third century A.D. Porphyry's *Isagoge* was a study of Aristotle's *Categories*—genus, species, difference, property, and accident—and it also posed the perplexing question of universals, one that still divides realists and nominalists in the twentieth century.[1] Anne Fremantle sums up the puzzle in the following way:

> This problem of Universals is: does whiteness, or fatness, or roundness, or anything that can be predicated of a number of things, really exist separately, apart from white things or fat things or round things? Or are the whiteness, the fatness, the roundness only in the things that are white and fat and round? Or are they only in our mind? Is whiteness merely a mental idea given us by all the white things we have ever known, and fatness the sum of all the fat things, and roundness the glomerate impression of all the round things we have perceived? If your reply is that whiteness exists, quite separately and substantially, then philosophically you are a realist; if you believe whiteness is only the mind's idea of the sum of things that are white, then in philosophy you are called a nominalist. (1954, 20)

An even more passionately disputed question, testing then-current ecclesiastical orthodoxies and still testing contemporary ideologies, is that of identity and definition. Thomas Aquinas earned condemnation by the (Platonic) authorities of his day for claiming that the soul was the substantial form of the body. On the individuality of human persons and the primacy of existence over essence, Thomas offered the following argument:

> Now, whatever belongs to a thing is either caused by the principles of its nature, as the ability to laugh in man, or comes to it from some extrinsic principle, as light in the air from the influence of the sun. But it cannot be that the existence of a thing is caused by the form or quiddity of that thing—I say caused as by an efficient cause—because then something would be its own cause, and would bring itself into existence, which is impossible. It is therefore necessary that every such thing, the existence of which is other than its nature, have its existence from some other thing. And because every thing which exists by virtue of another is led

> back, as to its first cause, to that which exists by virtue of itself, it is necessary that there be some thing which is the cause of the existence of all things because it is existence alone. Otherwise, there would be an infinite regress among causes, since every thing which is not existence alone has a cause of its existence, as has been said. It is clear, therefore, that an intelligence is form and existence, and that it has existence from the First Being, which is existence alone. And this is the First Cause, which is God.
>
> Now everything which receives something from another is in potency with respect to what it receives, and what is received into it is its act. It is necessary therefore that the quiddity itself or the form, which is the intelligence, be in potency with respect to the existence which it receives from God; and this existence is received as an act. It is in this way that potency and act are found in the intelligences, but not form and matter, unless equivocally. (Aquinas 1253, bk. 1, chap. 3, par. 80–81)

William of Ockham applied his nominalist "razor" to the basically realist assumptions on which Aquinas built his solution, thus also foreshadowing that persistent antinomy between rational (Latin/Continental) and empiricist (Anglo/American) preferences in scientific inquiry. Questions of existence and identity remain hotly debated in the twentieth century, and once again ontological arguments are entertained.[2]

These so-called Dark Ages of feudal social order and scholastic treatises on form were by no means as somnolent as later historians would paint them. Here was a continent innocent of national boundaries, across which scholars, missionaries, crusaders, and armies trekked. It witnessed barbarism and massive environmental changes, but out of the crucible of bloody interactions between north and south, Christendom and Islam, Roman and Goth, emerged the Renaissance Phoenix that would bring about profound transformations in world-view, power, and politics (see Chapter 2). The most fruitful legacies of medieval times for geography came from the Arab world and from that interaction of Muslim and Christian scholarship which became possible during the twelfth century. Pilgrims to holy sites, Christian and Muslim, assembled voluminous topographical information, writing itineraries providing much detail on the lands traversed,—for example, Marco Polo and Ibn Battuta. Roger II, Norman king of Sicily in the mid-twelfth century, ranks as one of the greatest sponsors of geographical research in the Middle Ages. In his service, Al-Idrisi prepared his *Geography* of the world and a celestial planisphere and round world map engraved in silver: a remarkable contrast to the flat-earth, Jerusalem-centered, maps of Christendom in that day (figs. 21, 22).

The single most eloquent expression of medieval formism is the Gothic cathedral: an edifice in which there was literally a place for everything and

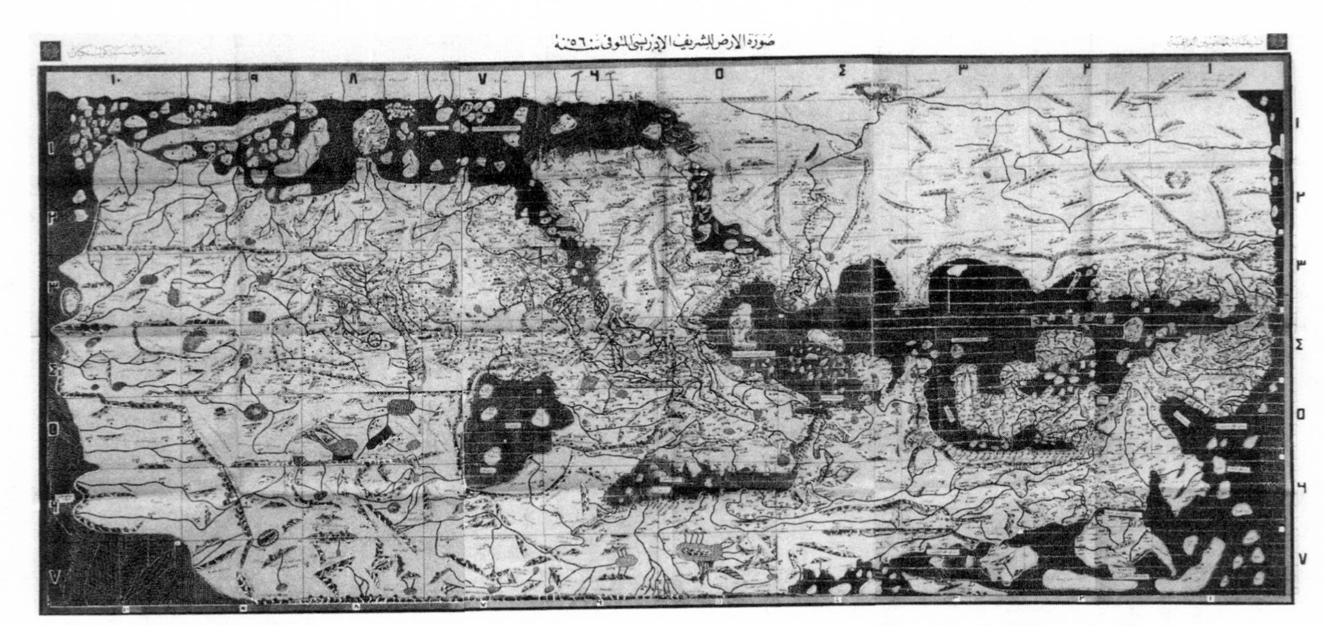

Fig. 21. Zonal map of the world based on Al-Idrisi 1154. (From the archives of Lund University. Photo by Bertram Broberg.)

everybody, from monarch to peasant, from craftsman to scholar, each with a patron saint or angel to guarantee correspondence between forms on earth and those in heaven. The entire orchestra pointed heavenward: form and matter embraced, as it were, in a giant cone, whose apex was God the Creator, First Cause of all existence (D. Jacobs 1969). "A comparison with cathedrals has come to many Western minds in searching for a metaphor for icebergs," Barry Lopez wrote, "and I think the reasons for it are deeper than the obvious appropriateness of line and scale. It has to do with our passion for light" (1986, 222).

> Cathedral architecture signaled a quantum leap forward in European civilization. The gothic cathedral churches, with their broad bays of sunshine, flying buttresses that let windows rise where once there had been stone in the walls, and harmonious interiors—this "architecture of light" was a monument to a newly created theology. . . . the erection of the cathedrals was the last wild stride European man made before falling back into the confines of his intellect. (222–24)

Equally eloquent is the Gregorian liturgy, whose content, rhythms, melodies, and forms celebrated a consensual affirmation of medieval Christianity's world-view, its sanctoral cycle attuned to the seasons of agrarian life (Mahrt 1979).

Renaissance Formism

If the architectural forms of monastery and cathedral articulated the central *Zeitgeist* of medieval Christendom, so would the villas and castles, observatories and academies of Renaissance days articulate the ethos of an age of adventure spanning the fourteenth through the seventeenth centuries. "Man as Measure," and human reason as ultimate source of truth and being, were slogans to announce humanity's will to abrogate the role of creator and first cause. Similarities and differences among people and landscapes would no longer be construed primarily in terms of divine providence; rather, they offered a challenge for the accumulation of knowledge for its own sake. With the rise of national consciousness and the expansive thrust of some European nations, questions of social class, privilege, and prestige would arise. Capital accumulation and proprietary claims to territory and resources kindled passionate and competitive commitment to innovations in technology, rhetoric, and art (Cosgrove 1984).

Renaissance formism, with its novelties in landscape, art, and architecture, could indeed be read as a Faustian feat of symbolic and material transformations of secular rather than ecclesiastical motivation. The artifactual

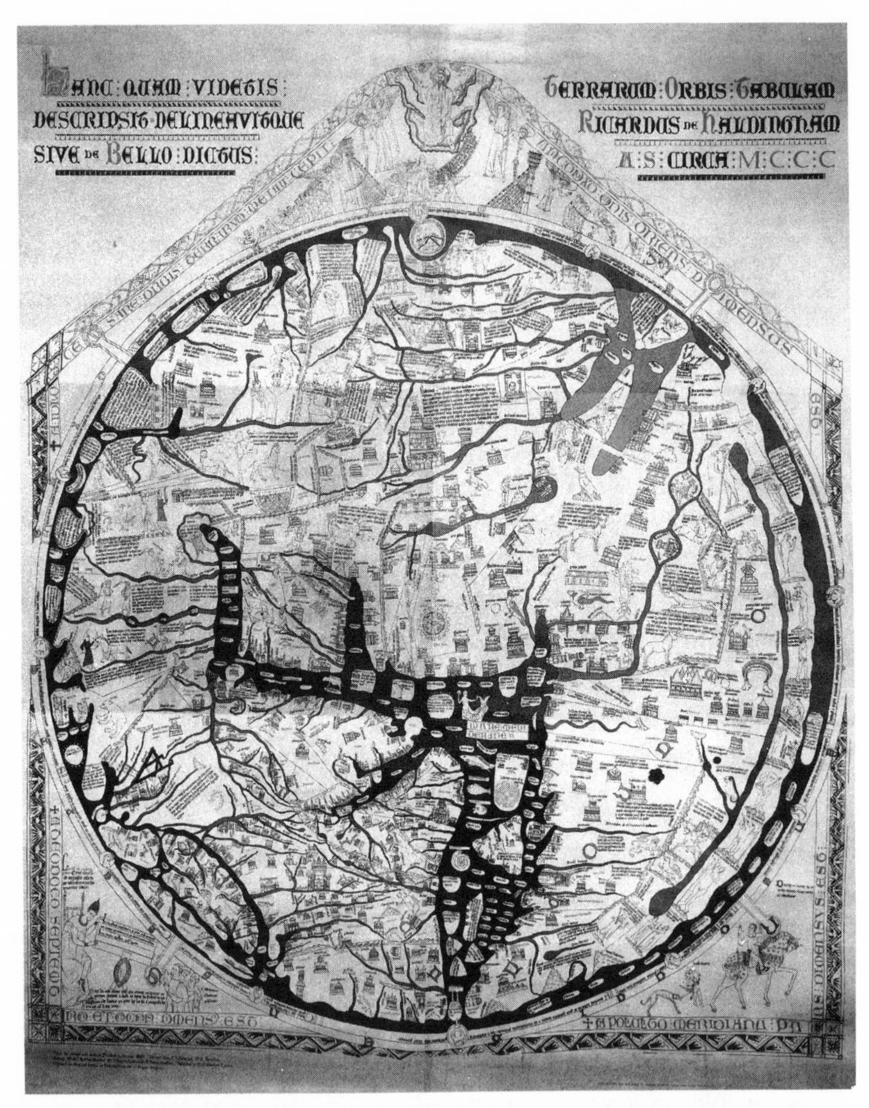

Fig. 22. *Mappa Mundi,* ca. 1500. (Photo reproduced by permission of the Dean and Chapter of Hereford Cathedral.)

landscape became a monumental expression of human creativity. Built forms reflected the social structures and power relations characteristic of emerging modernism. Vitruvius Pollio's ideal city incarnated the values of the emerging bourgeoisie, its forms corresponding appropriately to an anthropocentric rather than a theocentric artisanry:

> The ideal city is conceived as a unitary space, an architectural totality, a changeless and perfect form. It is delineated by a fortified wall, circular or polygonal in shape. At the centre is a large open space surrounded by key administrative buildings: the prince's palace, the justice building, the main church, generally referred to in classical terms as a temple, and often a prison, treasury and military garrison. . . . The dimensions of the central piazza and its architecture are rigorously controlled and strictly proportioned. Road patterns are determined by centre and periphery, either orthogonally or in gridiron pattern. They are designed to provide visual corridors giving prospects on key urban buildings or monuments. . . . Individual structures are to be designed according to the rules of the classical orders, thus each is rendered a microcosm of the same geometrical principles which govern the harmony of the whole city and which are displayed in the physical and intellectual properties of its citizens. The ideal city is designed for the exercise of administration and justice, for the civic life, rather than for production or exchange. (Cosgrove 1984, 94)

From a humanist vantage point, Renaissance geographical thought posed a paradox. Geographers themselves showed an ambivalence between two objectives of classical times, now subsumable under the rubrics of "Columbus" and "Ptolemy" (Broc 1986). On the one hand, there was much new empirical information emerging from voyages of exploration; on the other, there were horizons for cartography and mathematical geography, inspired by the rediscovery of *mappae mundi*. The former appealed mainly to adventurers, merchants, missionaries, and navigators, whose accounts (frequently circulated in vernacular languages) held enormous popular appeal but scarcely distracted the bona fide humanist from his Latin and Greek texts (Broc 1986). The latter led to innovations in mathematics and astronomy and theoretical perspectives on the world as a whole, which challenged the established authorities and orthodoxies of Roman or Alexandrian-based maps. The Renaissance scholar, unfettered from medieval orthodoxies, was thus confronted with a paradoxical situation. Astronomy and cosmography were revealing an image of the earth as just one relatively minute part of a larger planetary system. From exploratory voyages, too, there was undeniable evidence that the human world was vaster and more culturally diverse than even the Greeks had imagined. Some Renaissance humanists ignored both trends (de Dainville 1941; Broc 1986). Narcissus-like, several focused on challenges at home, thoroughly fascinated with prospects of resuscitating classical lore, serving the military and political interests of their sponsors; others created marvels of landscape, art, and architecture or pushed the limits of orthodoxy on matters of religion.

The rediscovery of Ptolemy's map and Pliny's natural history heralded a Phoenix mood, pushing imaginations to explore horizons beyond the geocentric orthodoxy of Mediterranean Christendom. Formism would be challenged by other equally plausible ways of explaining reality—for example, mechanism and organicism (see chapters 4, 5). Speculations about natural law, hitherto couched in terms of the *terrae incognitae,* inevitably clashed with notions of environmental determinism when one sought to interpret *terrae incognitae* in terms comparable to one's own world. A vast literature developed around the question of the optimal (temperate) climate (Leboulaye 1975–79; Huntington 1945, 343–67). Northwest Europe replaced Greece as having the ideal climate, and other societies were stereotyped in terms of their ability to lead a healthy life (Glacken 1967). The consistent curiosity about social forms and appropriate government for people with different temperaments, inhabitants of different climatic milieus, was fascinating. Bodin usually concluded his essays with a plea for a strong central monarchy as the only means of dealing with the schisms and strife of sixteenth-century France (Bodin 1606).

The Oriental world meanwhile entertained a similar curiosity about sociopolitical forms and natural environments. In 1773, Master Jiun preached that "in this world there are the true laws which benefit it always. Those who have open eyes can see these Laws as clearly as they see the sun and moon. Whether a Buddha appears or whether a Buddha does not appear this world exists and human beings exist. These Ten Virtues will always be manifest as long as they exist" (Atkinson 1905).

Parallels have been drawn between the work of Jiun and de Groot (1583–1645), the distinction being that the latter believed in God as author of natural law, whereas the former held that nature and law were nothing but Buddha himself (Atkinson 1905; Kinami 1973, 41). Huan Tsung-hsi (1610–95), a Confucian thinker, stressed the importance of governmental form rather than simply the character of the men administering it. "If men [who govern] were of the right kind," he claimed, "the full intent of the law would be fulfilled; and even if they were of the wrong kind, it would be impossible for them to govern tyrannically and make the people suffer. Therefore I say we must first have laws which govern well and later we shall have men who govern well" (Spea 1948, 205). There is a distinct parallel to the Western theology of the sacraments, whose power was independent of the worthiness of the minister (*ex opere operato*).

Transcendent formism thus regarded the visible patterns of the world in terms of underlying norms or ordering principles, be they conceived as *natural law* in the West, *rta* in the Rig Veda, *moira* in Greek, *Tao* in Chinese, or *-michi* in Japanese (Groot 1910, 12; Davids 1881, 278–89). The world seen as mosaic of forms, with a place for everything and everything in its place, has been a favored root metaphor in many of the world's civilizations, and echoes can be found in twentieth-century views such as Heidegger's—for example, in his essential conception of "dwelling" (1954, 1971).

Enlightenment Geography

While the intellectual imaginations of seventeenth- and eighteenth-century European scholars were stimulated toward Copernican views of the universe, and eventually Cartesian and Newtonian views of science, many geographers apparently clung to a formistic world-view, busying themselves with massive compilations of exotic data, or with perfecting the art of cartography. If in Renaissance times they could be regarded as "anatomists of the great world" (Crooke 1615), by the seventeenth and eighteenth centuries they were to become its gazetteers. With the rise of nation-states and the centralization of power after Louis XIV (1638–1715) and Frederick the Great (1712–86), there was demand for statistical surveys, for inventories of resources, population, and agricultural land. Geographers, together with political economists, engaged in practically oriented tasks, spurred on by the spirit of the *Encyclopédistes*. Publications such as those of Süssmilch and Büsching in German, Malte-Brun in French, and William Petty in English showed a concern about demographic and economic patterns, while Linnaeus (1707–78) and his disciples devoted more attention to botanical and economically useful inventories (see Frängsmyr 1983, 1989). Innovations in mapping and survey skills created a virtually unquenchable thirst for information, classification, and compilation, a phenomenon that would recur in the late twentieth century.

Bernhard Varens (1622–50), commonly known as Varenius, stands out as the central transitional figure between Renaissance and Enlightenment geography. "Geography is called a mixed mathematical science which teaches the properties of the Earth and its parts as far as they can be measured, namely: its configuration, locality, size, movement, celestial phenomena, and other properties" (Varenius 1672). Varenius distinguished between *general geography,* namely, study of the earth as a whole, and *special geography,* describing particular regions (chorography) or localities (topography).[3] There were three major principles to be used by geography in proving its propositions: (1) the propositions of geometry, arithmetic, and trigonometry; (2) the precepts and theorems of astronomy; (3) experience and the observations of people who described the individual regions.

This framework retained indubitable elements of classical formism, its logic and physics thoroughly Aristotelian, but its world-view Copernican rather than Ptolemaic. It bore within it elements of two very distinct "world hypotheses," to use Pepper's phrase, formism and mechanism. Isaac Newton recognized its value as an empirical grounding for a mechanical interpretation of the universe. The most enduring legacy of Varenius's theory was the credibility it afforded to formism as a foundation for the discipline of geography. This was most probably due to Kant's reading of Varenius (May 1970). In the introduction to Kant's *Physische Geographie* (1802), as well as to Hettner's *Die Geographie: ihre Geschichte, ihr Wesen und ihre Methoden* (1927), the Varenian framework was cited; propositions (1) and (3) were reiterated, but (2)—the principle that was closest to Newtonian—was omitted.

Immanuel Kant (1724–1804), a towering figure in Western philosophy, taught geography at Königsberg for thirty years. In his elaborate systematization of human knowledge, he assigned a place for geography and epistemological guidelines for the discipline which have consistently appealed to leading practitioners to this day. Formism, as world-view, owes its enduring hold on the geographer's consciousness to the epistemological guidelines set down by Kant: chorology, morphology, spatial analysis, and landscape interpretation would henceforth be regarded as indubitably "geographic" pursuits.

If issues of authority claimed overriding importance during medieval and Renaissance times, those of cognitive certainty would be far more compelling for Enlightenment scholars. Kant posed four major questions: What can I know? (metaphysics), What ought I to do? (morals), What may I hope? (religion), What is Man? (anthropology). On the purely cognitive aspects of such questions, however, Kant confronted the two opposing stances of his day: was knowledge attainable through the exercise of reason (as Plato, Thomas Aquinas, and Descartes believed) or through the evidence of the senses (as Democritus, William of Ockham, and later Hume and Locke believed)? He assumed a position that he hoped could transcend this dualism. He began by affirming the rational stance: Descartes' distinction between mind (*res cogitans*) and matter (*res extensa*) was transposed to the distinction between *noumena* (thought constructs) and *phenomena* (things in the world). One could only be sure of what went on in the head: mind would be the source of order. One could never be sure of causality on the basis of empirical observations alone; it was the intellect that engaged in the process of intuiting properties to the real world (I. Kant 1919; May 1970).

Yet Kant also acknowledged the need for empirical observation and hence the value of geography. Like Euclid, Descartes, and Newton, he regarded time and space as absolute properties of the world, a priori conditions for human intuition. Time, the inner sense, was seen as seeking order and explanation; space, the outer sense, as constantly generating diversity. Then there was the third sense, namely, causality (essence, number, moral meaning), and this would seek explanation. The model for science, that is, objective explanations of things, was without doubt Newtonian physics and mathematics. In the exercise of theoretical or pure reason, the "transcendental subject" could overcome the apparent conflict of inner and outer senses, thereby achieving objectivity. For practical reason, however, for example, in ethics and moral questions, there were no such objective ("object-oriented") foci; in this context, the objects were other people. In the exercise of practical reason, therefore, Kant appealed to the famous *categorical imperative*—a belief that all humans, attuned to the voice of reason, would eventually agree on a common code of moral behavior. To achieve "transcendence" in the exercise of practical reason, the subject would need self-knowledge, self-control, and self-realization.

In Kant's comprehensive and systematic classification of the sciences, however, geography would be assigned a basically descriptive role (May 1970):

SCIENCES	**SPACE**	**TIME**
	Outer sense	
Empirical	Geography	Anthropology
(descriptive)	History of Nature	History of Morals
Theoretical	Physics	Psychology
(explanatory)	Other Natural Sciences	

Geography for Kant qualifies as a gentleman's essential education. It provides "a purposeful arrangement of our perceptions, gives us pleasure and provides much material for friendly discussion" (May 1970). Taken literally, Kantian geography could only be a descriptive body of study: pedagogically speaking, it was an excellent and essential *propedaetique;* as scholarly craft, a field for mapmaking. On Kantian foundations geography could become a chorological science, as later advocates of the so-called spatial tradition would reaffirm, especially those of Idealist orientation such as Hettner, Penck, De Geer, van Paassen, and Hartshorne. But Kant had also emphasized the essential relationships between history and geography. The "general outline of nature," which physical geography provided, was not only the basis for history but also the foundation for other possible geographies. He announced a prospectus for mathematical, moral, political, commercial, and theological geography. In all of these, formal patterns could be seen as expressions of underlying norms, eventually causes, which would constitute the foci of other systematic sciences. Strictly speaking, Kantian geography, as an empirical science, could yield only two kinds of product: a conceptually based classification of phenomena (e.g., the Linnaean survey) or a cartographic representation of the distribution of phenomena in space. Fortunately for geography, many of its formists have stretched the limits of Kantian orthodoxy and have produced classics of regional description such as those of *la géographie humaine,* of the Berkeley School, of sequent occupance, and of Domesday historical geography.

Kantian Foundations and the Chorological Imperative

Carl Ritter's *Erdkunde* (1862), written in 1815, and often regarded as the foundational text for academic geography, is Kantian in spirit and offers a good example of transcendent formism (Ritter 1817). Two major features of this work rank it as far more sophisticated than the gazetteering accounts of the previous century. First, observable data are integrated around the notion of *region,* a framework within which unity in diversity can be discerned. Second, there is the element of *diachronic flow,* an identifiable historical process that transcends all the apparent forms and regional *Gestalten.* Geography could become a study of the terrestrial unfolding of a divine plan for humanity. Ritter's influence radiated widely: the *Erdkunde* held enormous appeal in schools and academies throughout Europe, and eventually, via Arnold Guyot, in America.

On the eve of disciplinary establishment, there were other root metaphors in the air, and formism would now and then yield to the apparently more integrated approaches heralded by mechanism and organicism. Formism offered a convincing solution to some of the first major impasses faced by the discipline in Germany—for example, how could one maintain both physical and cultural aspects within one single field and at the same time claim status as a science? Better to separate them, some (such as Gerland and Fröbel) argued, for apart from the difficulties of incorporating human and physical subjects within a single scientific approach, the prospect of studying relationships between humanity and its environment was clouded with fears about determinism (Van Valkenburg 1951; Dickinson 1969). Ratzel's *Anthropogeographie* (1882–91) offered one solution to this impasse, a solution appealing to organicists but thoroughly unwelcome to formists. The Ritterian solution of synthesis by regions, as well as his theologically inspired integration of historical and geographical perspectives toward a global view of process, was also unpalatable to materialists. Alfred Hettner, in a series of influential articles written from the end of the nineteenth through the early twentieth centuries, successfully reiterated a formist stance on the nature of geography.

Hettner claimed that geography could be regarded neither as a science, in the sense of *Naturwissenschaft,* nor as a field of the humanities, in the sense of *Geisteswissenschaften*. He was convinced that both physical and human components should be included in the practice of the discipline, without, however, assuming an environmental-determinist stance. Geography could not be defined in terms of region, landscape, or any other special category of phenomena; that is, it could not claim these as its own special domain. How, then, might one locate a core substantive focus for the field? For him, neither post-Darwinian "ecological" perspectives nor Hegelian-type teleological arguments about "unity of nature" could provide an ideologically acceptable foundation. He would define geography as a "chorological science of the earth's surface" (Hettner 1927). Echoing Kant, he distinguished between sciences of "how" (process-oriented scientific inquiry), of "when" (time-oriented chronological tracings of things), and of "where" (sciences of spatial relations). Geography would henceforth be identified in terms of its special (spatial) perspective on reality: distributions, areal differentiation, location, and pattern.

The welcome with which Hettner's formulation was greeted in most major schools of geography has much to do with the empirical realities and political events of the pre– and post–World War I era. A formist solution was attractive on ideological and emotional as much as on epistemological grounds: it seemed to provide a neutral ground on which to base international cooperation within the discipline. At the Ninth International Congress of the International Geographical Union, held in Geneva (1908), the following motion was adopted:

> Geography . . . has as its object the description of the surface of the earth, considered in its various elements, physical and living, whose combination and interdependence determine the existing physiognomy of the globe. This teaching in primary and secondary education should be based on the reading of maps and should pursue above all a synthetic method. In the portrayal of the different parts of the globe, geography should make clear the relation between the inorganic world and living

things, and more particularly between the surface of the earth and man. (Cited in Dickinson 1969, 270–71)

During the interwar period, Hettnerian formism (morphology and chorology) was defended and practiced with enthusiasm on both sides of the Atlantic (De Geer 1908, 1923; Baker 1917; Hartshorne 1939; Darby 1947; de Jong 1955). Formism could invite practically oriented empirical work, as well as a virtual dismissal of such philosophical questions as had preoccupied the earlier generation. Fundamental differences of cultural and intellectual traditions, however, meant that the terms *morphology* and *chorology* were applied to very different modes of research. Hettner's approach to *Kulturlandschaft,* a model intimately associated with the historical realities of central European life and milieus, was not easy to transpose to a midwestern American context (Hettner 1905, 1927). Besides, as Butzer has recently pointed out, the conventional tendency to identify Hettnerian geography with regional description (and areal differentiation) reflects some ignorance of Hettner's own stances on process, explanation, and genetic geomorphology (Butzer 1989). For a great number of the interwar generation, "explore and map" was sufficient challenge and was considered suitable for practitioners of both human and physical branches of the field. Geomorphology, in its sibling rivalry with geology, could debate the priorities to be given to "form" and "process," formists readily admitting that systems-analytical approaches could offer explanation of process, but that their *explicanda* would always be pattern and form; mechanists and organicists would reverse the priorities (fig. 23).

In human geography, this debate over formal versus functional interpretations of life and landscapes was delayed for a generation or so. Besides, up to World War II there were very few proponents of a "process" orientation to demographic, economic, or political patterns. Each country experienced its own controversy about what geography ought to be during the twentieth century. For brevity's sake, a focus on the North American story will suffice here to illustrate the strengths and limitations of formism in the changing contexts of its performance.

Mosaic and American Geography

Formism held a particular attraction for North American geographers in the wake of World War I. Environmental determinism, which had characterized education for at least two generations, was something the younger recruits to the discipline wished to abandon. They also felt the need to climb out from under the umbrella of European masters, and were obviously uneasy about the "mystical" verbiage and sometimes disconcerting

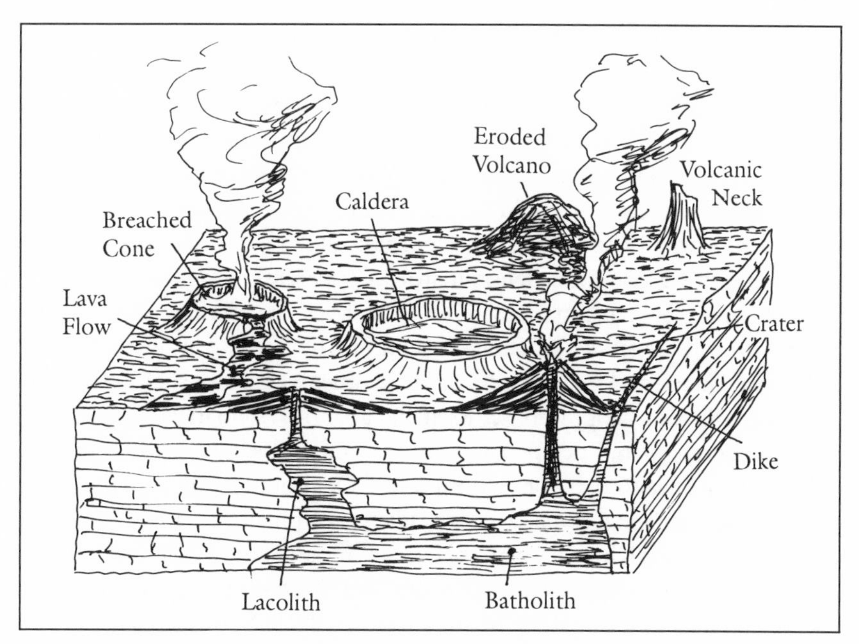

Fig. 23. Form and process. (Sketch by Bertram Broberg.)

implications of an organicist world-view. Colleagues in history were already moving away from *frontier* theses to analyzes of *sections* (Turner 1920, 1961); colleagues in sociology were moving from the study of "natural areas" in Chicago to "social area analysis" and the sociometry of demographic patterns (Theodorson 1958). Geography, as the study of space and areal associations, could still maintain a house of many mansions and, besides, become visible and useful within society at large (PG, 62–65). Robert Platt reminisced: "In comparison with my philosophy major, geography offered the advantage of dealing with tangible and visible things forming a solid basis on which to build ideas, instead of beginning and ending with abstractions. In comparison with my history major, geography had the advantage of going more to the field for direct observation instead of going to the library to read about things no longer visible. In comparison with geology, geography had the appeal of dealing with the world of people instead of only rocks and fossils" (cited in Hartshorne 1964).

Three features of the American esprit during the interwar period which made formism attractive at that moment are obvious: the appeal of concrete, tangible objects of analysis rather than abstractions; the primacy of present over past; and the superior value of empirical and practical problem solving over laboratory research (Blouet 1981; James and Martin 1979). Preston James, reflecting on the founders of American geography, extolled the chorological approach, for example, Powell's attempts to persuade Congress to reconsider its commitment to a rectangular grid for new settlers in the West; Johnson's evidence for the feasibility of the Panama Canal; and even George Perkins Marsh's warnings about soil erosion in the Mediterranean. In every case, it could be shown how valuable the "map" was in elucidating complex issues (Marsh 1864; Powell 1878; P. E. James 1981).

Even with a common denominator as apparently tangible as form or pattern, however, tensions inevitably arose over the identity of the discipline within the overall academic division of labor. A few individuals, such as Wright, Sauer, and Hartshorne, did take time to consider such theoretical questions. Should geography set itself the goal of systematically presenting its own special spatial perspective on phenomena such as settlements, circulation routes, land uses, or cultural traits? Or should it cultivate the art of piecing together the *Gestalt* (*Zusammenhang*) of phenomena within particular regions? Resolving the perennial issue of parts and wholes, of analysis and synthesis, became a virtually insurmountable task, given the masses of specialized information which by then were available from various branches of chorological inquiry. John K. Wright would transpose this *problématique* onto another level of discourse, steering imaginations toward a radically different world-view, one closer in many ways to the American spirit, namely, toward contextualism (see Chapter 6). The majority of geographers breezed along with empirical (descriptive or problem-oriented) tasks, rarely seeing any point in critical reflection about theories of science. Hartshorne sufficed.

In the writings of Hettner and Hartshorne, the implicit or explicit bibles of interwar practice, there was encouragement for both systematic and regional perspectives; they should ideally complement each other, as long as both retained a commitment to elucidating landscapes. This doctrine had quite a different echo among chorologists on the eastern and western sides of the Rockies, however. Carl Sauer and the Berkeley School maintained a sensitivity to history and cultural diversity, focusing on material forms as tracers of cultural and natural history, opposed to evolutionary or deterministic theories of history (Sauer 1941; Speth 1981; Solot 1986), and sneering occasionally at the functionalist preoccupations of their former colleagues in the Midwest (Rostlund 1962; Leighly in PG, 80–89; Butzer 1989). There was another distinct cleavage, however, between those who, following in the spirit of Ptolemy and Ortelius and armed with positivist epistemology, eventually sought to model and theorize spatial patterns and process, and those who, with historicist and idealist presuppositions, articulated the uniqueness and individuality of places and regions. The former, eventually rejoicing in their creative encounter with another world-view, namely, mechanism, played a key role in engineering a "New Geography" after World War II; the latter would open their doors to fresh philosophical currents emerging from postwar Europe—for example, hermeneutics, critical theory, existentialism, and phenomenology.

In the wake of World War II, formism hibernated for awhile, wartime experiences having engendered a powerful trend toward mechanistic interpretations of reality (PG, 186–96). The actual experience of war had transformed that world-view, which had dominated the discipline for at least a generation with its tidy housekeeping methods, geometric forms, and slumbering regions. The inductive style of field research soon yielded ground to the deductive, nomothetic aims rivaling the ideographic. "Processes of spatial interaction" (Ullman 1954) competed with "patterns of areal differentiation" (Hartshorne 1939, 1959) as key slogans for the discipline. Externally, too, the conquest of the air and aviation technology brought about a radical change in conceptions of space and time. An areally differentiated world became transposed to a topological surface of points, lines of force, and shrinkable or expandable distance. Postwar reconstruction plans in virtually every Western society again kindled the Faustian dream: via science and technology, rationality would guard against future war; rationalization of space and functions would harmonize criteria of efficiency and order. Geographers would begin to emphasize process rather than pattern, function rather than form, and some would explore their interactions.

It was just this challenge of linking form to process that gave rise to the midcentury "reformation" in Anglo-American geography. Beginning at Iowa in the 1950s and quickly linking to Seattle, Evanston, Chicago, and Michigan, a quantitative wave, anchored in positivist methodology, gave formism a revolutionary facelift. Bunge's *Theoretical Geography* (1962) echoed the Pythagorean dream: chorology could be pursued with geometric rigor; geography could scoff at its (Hettnerian) exceptionalism and become a science again. "Locational analysis" became a kind of methodological flagstaff heralding prospects for geography's analytical potential in a wide variety of domains, from dynamics of river basins to social patterns of cities, location of services, and diffusion of diseases (Chorley and Haggett 1967; Harvey 1969; Haggett 1972; Amadeo and Golledge 1975). In the late 1950s and early 1960s, Sweden became a kind of model for scientific and applied geography much applauded by advocates of the reformation.

By the late 1960s, a number of sources had begun to criticize the opaqueness of such representations of reality as were emerging from the quantitative school, the limitations of positivist procedures, and the value implications of plans for urban or regional development which were emerging from regional science. Though ideologically diverse, much of this critique was conducted in the categories of formism: one spoke about tensions between "lived space" and "representational space," between "insiders" and "outsiders" and their different "mental maps," eventually between their radically contrasting experiences of space and place. In many of the alternative approaches sampled in the sixties, the aim was primarily descriptive or heuristic, and their conceptual and methodological challenges were those of "immanent formism," namely, classification, categorization, and symbolic representation. One sought to unmask the isomorphism (or lack of it) between distributional patterns, for example, examining the "fit" of various socio-demographic spaces to existing boundaries of administrative and electoral domains, or the location of services. Multivariate analytical techniques, for example, factorial ecology, yielded pictures of various "spaces" (fig. 24).

Certain norms could be suggested from such descriptions and representations—for example, the consistently inferior status and spatial segregation of certain groups (Herbert and Johnston 1978; Jones 1986). Immanent for-

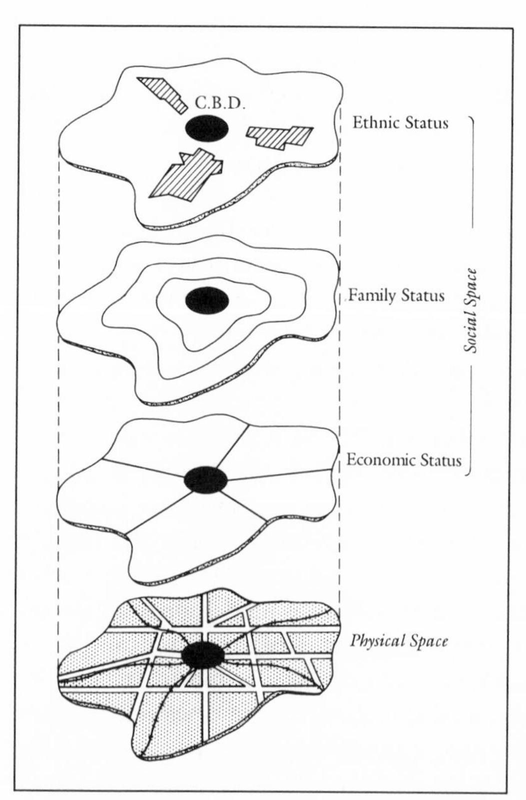

Fig. 24. Levels of social space. (Reproduction courtesy of Robert A. Murdie.)

mism sufficed to document inequalities, injustices, and inefficiencies, but it lacked explanatory theory and was inadequate *in se* as basis for normative action. Only later, when it was combined with mechanistic models of explanation, did any integrated theory emerge. In fact, few latter twentieth-century geographers could focus exclusively on form itself, without reference to underlying process.

Other trends, however, had more ambitious goals in sight, more akin to transcendent formism. At one extreme, the individual came into focus—for example, through phenomenology, where one explored the meanings of space, nature, and environment for human experience, seeking, in Husserl's famous phrase, to "go back to the things themselves" and let reality speak for itself (Tuan 1971; Samuels 1971; Relph 1974; Seamon 1980). At the other end of the spectrum, one sought structural explanations for the forms and patterns of space and society—for example, through the theoretical lens provided by Marxism and critical theory (Lefèbvre 1974; Harvey 1972; Santos 1975; Peet 1977). An existentialist motif characterized many of the pioneering voices of these two movements: the antinomy of subject and object, of folk and elite, of individual versus managerial interests, were all emphasized. Implicit norms guiding the diagnosis and potential solution of problems were, however, radically different. While the former would seek exlanations and practical implications in human nature, sociality, and grass-roots *consciençizacão,* the latter sought them in transcendent structures and their transformations. There were short-lived moments of creative encounter between these streams of postbehavioral geography in North America; sustained dialogue between the two would happen elsewhere, among people more closely attuned to Continental philosophical traditions and in lands of Latin cultural background (Santos 1975; Olsson 1979; Racine 1977; Pascon 1979; Gomez-Mendoza and Cantero 1992).

The vocational appeal of *ergon* was certainly not universally shared by late twentieth-century geographers. The *poetic* appeal of form and pattern not only led to renewed interest in historical and cultural geography, landscape appreciation, and livelier interaction with archaeologists (Butzer 1978); it also rekindled research interest in the sense of place, territorial identity, and human creativity (Buttimer and Seamon 1980; Jager 1975; Richardson 1982; Seamon and Mugerauer 1985). For many, the cultural landscape became a fascinating focus for the testing of diverse hypotheses about society, power, ways of life, literature, and landscape (Pocock 1981; Cosgrove 1984; Daniels 1988; Rowntree 1988). In these research directions, formism often converged with contextualism, a world-view that regarded the world as a possibly unique arena of events, each of which should be elucidated in its own terms (Chapter 6). Both of these world-views shunned globally integrated theories about reality; their proponents, especially those of humanist orientation, endeavored as far as possible to allow empirically documented patterns to "speak for themselves." In terms of *logos,* the contextual sensitivity generated during the 1960s, for example, in perception research, implied for some a virtual redefinition of the researcher's own stance—from that of observer to that of participant—one outcome of which was a more hermeneutic approach to landscape interpretation (Samuels 1971; Meinig 1979; Rose 1981). "Our human landscape," Pierce Lewis remarked, "is our unwitting autobiography, and all our cultural warts and blemishes, our ordinary day-to-day qualities, are there for anybody who knows how to look for them" (1979, 13). What geography had lost, in terms of *poesis* and *paideia,* in the battles of certainty and uncertainty which post-war generations had conducted, was recaptured in the discovery of meaning and order in landscape, and in the artistic challenge of communicating such knowledge effectively (Lewis 1986).

The flowering of humanistic interest in various branches of geographic endeavor during the 1970s brought confusion as well as joy. In stark contrast to the alleged certainties of locational analysis and regional science, to the well-intentioned efforts in applied geography by the socially conscientious, and to the Faustian energy expressed in creating new research fields and methodological innovations during the late 1960s, one was now in an era of relativism, contextual contingencies, and deconstruction. Coincidentally, of course, one was facing a new economic depression as far as funding for geographical research or for university departments was concerned. Not surprisingly, therefore, bold attempts have been made to rescue the discipline from the ambiguities and sometimes nihilistic implications of radical empiricism and a philosophy of science where "anything goes" (Feyerabend 1961, 1975). Some sought common ground in epistemology, language, or mathematics (Gould and Olsson 1982; Scott 1982); others assumed administrative responsibilities outside of geography, hoping to engineer better prospects for its disciplinary survival. But among the rank and file, new philosophical movements were astir which might hopefully restore some sense of security and promise for geography's ability to deliver "indubitable truth," or at least epistemologically defensible explanations of earth and world. From Continental sources, structuralism offered some attractive possibilities, and in its Anglo-American reformulations, realism again afforded appeal.

Robert Sack's *Conceptions of Space in Social Thought* laid out a "broadly conceived realist scientific framework to incorporate the multiple conceptions of space raised by our awareness of different viewpoints and philosophical approaches to human behavior" (1980b, 4). This was indeed a brave enterprise, and a bold stroke for reinstating a formist world-view in geography. Like medieval realists, Sack explored the ontological as well as the epistemological foundations for different ways of construing space (fig. 25).

Realism offered prospects for delivering a true picture of the world. Its aim was to isolate "natural laws," which operate independently of human experience, and culturally variable modes of scientific analysis (Bhaskar 1979). In the context of social science, one was urged to focus on questions

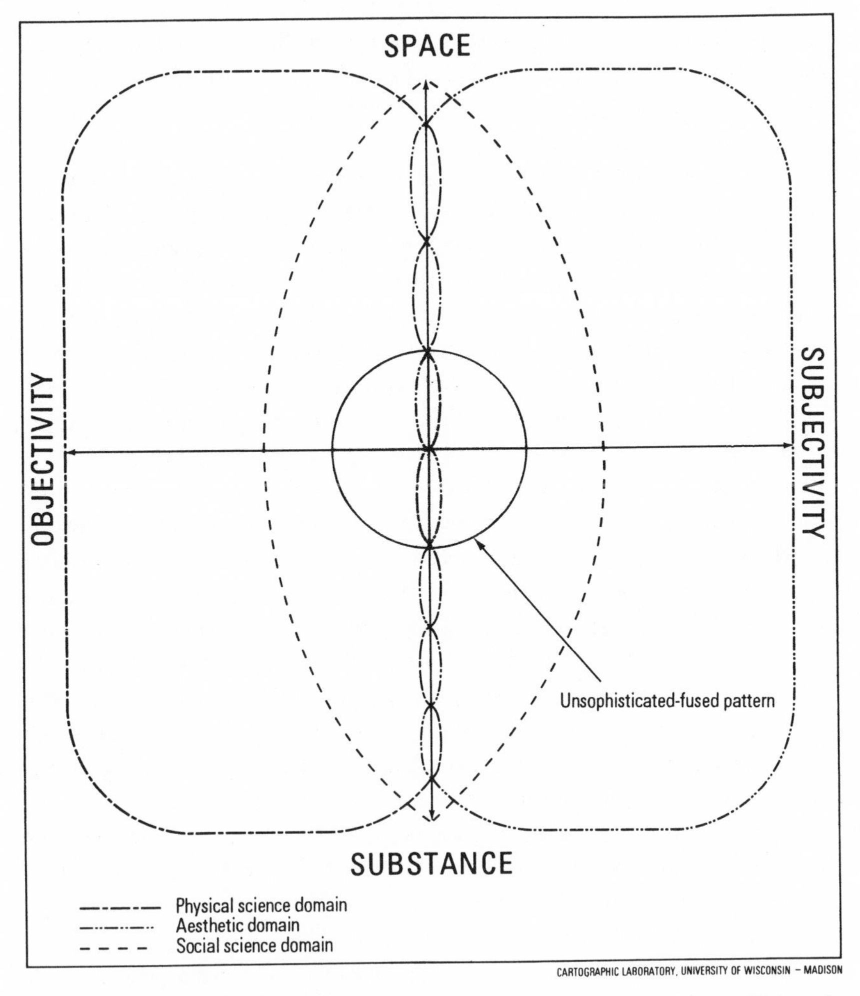

Fig. 25. Spatial schemata, after Sack 1980a. (Reproduced by permission of Edward Arnold Publishers.)

of "what" and "where," rather than on questions of "why" (Keat and Urry 1975; Urry 1981). In this, there is an easily recognizable echo of Whitehead's dream of neutralism for intellectual inquiry (Whitehead 1933) and Bertrand Russell's conviction that, after all, data perceived via the senses and screened through common sense could yield explanations of reality which were universally verifiable—for example, the laws of gravity, the ebb and flow of the tides, the regularities of temperature, insolation, and precipitation in the maturation of plants (Russell 1959). Realism holds significant appeal today for bona fide adherents to the empiricist and inductive canons of scientific inquiry once set forth so boldly by Locke, Hume, and Anglo authorities, as well as for those who still find meaning in the Kantian prospectus for geography (Sayer 1982; Johnston 1983; Lawson and Staeheli 1990). A step beyond positivism beckoned to seekers of certainty: whereas positivist procedures only promised to deliver truthful statements, realism promised "the truth" (van Fraassen 1980, 9).

Structuralism also sought to identify *constructs,* or "forms of universal mind," which could eventually be seen as transcending individual "minds." Here, one found postulates about "deep structures," or *archetypes* in collective consciousness, which articulate themselves in corresponding forms of language, social structure, and landscape. Scrutiny of these forms, it was claimed, could ultimately yield universal truth to transcend the peculiarities of individual consciousness (Jakobson and Halle 1956; Lévi-Strauss 1966). Within structuralism, as within realism, there are many genera and species. While some regarded forms as primary explicanda (immanent formism), others regarded process underlying form (transcendent formism) as far more exciting. The tensions between immanent and transcendent formism, as well as between formism and mechanism, have become dramatically apparent in recent structuralist discourse (Rose 1987; Gregory 1978, 1985). Tensions between idealist and materialist propensities among structuralists (often revealing differences between organicist and mechanist world-views) continue to cause debate. During the 1970s one heard much ado about "structuration theory" and the possibilities it afforded for including (if not always harmonizing) considerations of time as well as space, agency as well as structure, pragmatic relevance as well as the search for universal principles (Gregory 1985; Johnston 1983; Pred 1984). Not that radically different from the promises afforded by "transactional analysis" in sociology a decade or so earlier, these approaches ultimately pointed toward contextual rather than formist world-views (Bourdieu 1979; Giddens 1979).

As in Renaissance times, the compelling challenges for formism in the late twentieth century were to emerge from "secular" developments in technology and politics. An unprecedented abundance of information, from satellites or from more refined, computer-based information files, has once again raised the classical dilemmas of *geographia* and *historia:* how to interpret, explain, and use all the geo-information that now floods in around us. Like their humanist forebears in Renaissance times, many humanists simply turn away. Like their Renaissance forebears in commerce, evangelization, and navigation, too, many technicians forge ahead in creating new vocabularies, new vernaculars, for a universal information society. Erstwhile pioneers of geography's quantitative reformation, whether of mathematical or statistical bent, envision great possibilities for world communication networks, and, like the *Encyclopédistes* and scientific humanists of Enlightenment days, encouraged education for universal computer literacy on the grand prospects of tomorrow. Political authorities again offer sponsorship for formistic geographical analysis, but often only technicians appear qualified for the task.

The world as mosaic of forms, one of geography's most enduring world-views, is only one of many alternatives in Western history. It has danced elegantly at times, stumbled awkwardly at others, in settings where other choreographies won stronger appeal for their explanatory power. In Pepper's language, this dispersive world-view remains unlimited in scope but always threatened with lack of adequacy. It therefore seeks connections with mechanism in order to gain explanatory power. Illustrations of the many ways in which formism and mechanism complement each other can be found in many twentieth-century fields of science and humanities, especially biology (D. W. Thompson 1917; Judson 1979, 1980; Sheldrake 1981), geomorphology (Lidmar-Bergström 1984), and mathematics (Thom 1975; Mandelbrot 1983). Late twentieth-century developments in landscape ecology, information theory, and geographical information systems have once again revitalized scholarly interest in this root metaphor of reality (Phipps 1981a, 1981b; Phipps et al. 1986; Baudry 1989). The perennial puzzle of how form relates to function, pattern to process, is likely to offer a persistent challenge to geographers around the world. Peter Haggett perhaps reiterates the sentiments of many in the conclusion of *The Geographer's Art:*

> For me the basic puzzle and riddles of geographical structure remain enduring. The structural symmetry of the planet as viewed from outer space, the sequence of atolls in a Pacific island chain, or the terraced flights of irrigated fields on a Philippine hillside continue to be awesomely beautiful. When the time dimension is added, then changing structures take on added gleam. Diffusion waves become an interweaving dance of trajectories in multi-dimensional space. (1990, 184)

Chapter Four

World as Mechanical System

> The world is like a rare clock . . . where all things are so skillfully contrived, that the engine being once set a moving, all things proceed according to the artificer's first design, and the motions . . . do not require the peculiar interposing of the artificer, or any intelligent agent employed by him, but perform their function upon particular occasions, by virtue of the general and primitive contrivance of the whole engine.
>
> —Boyle 1690

IF PEGASUS, PAN, or the Platonic artisan are guiding symbols for a formist world-view, then Prometheus, who stole fire from the gods for the benefit of humanity, is surely the hero of a mechanist one. While formists sought a highly diversified and dispersed panorama of patterns, each elucidated in terms of its own nature or underlying norms, mechanists would explore processes of interaction, functional connections, and eventually causes for how the world works. One might associate formism with chamber music, its cadences and melodies echoing a world at rest, dormant and settled landscapes, inhabited by societies with a high degree of consensus and established order. In this vein, mechanism would be the strident bivouac of marching armies, the liberation songs of revolutionaries, or the clockwork beat of experimental gadgetry, factory, and locomotive. Formists, it has been claimed, tend to accept the status quo, to favor liberal and democratic politics, whereas mechanism has been the favored metaphor of revolutionaries and promoters of change in social and intellectual life (H. White 1973).

Generalizations of this kind invite awareness of the many exceptions. Nor should the contrasts of world-view between mosaic and mechanism imply that their cognitive goals are always mutually opposed. Distinct as they are, they have frequently converged in new explanatory models in physics, geography, biology, architecture, and the social sciences. Geographers of diverse backgrounds have puzzled over questions of landscape *morphogenesis,* probing ways in which forms on the earth's surface have resulted from physical processes, evolution of soil, plant, and animal life, and eventually the vicissitudes of human history. As they now seek to understand unprecedented changes in mankind's terrestrial home, scholars are rediscovering the complementarity of these two root metaphors.

Curiosity about functional relationships in nature, the human body, and ways in which the elements of fire, air, earth, and water could be harnessed for the practical necessities of life are as old as humanity itself. If the mandala offers a central symbol for the world as mosaic, then the wheel or level symbolizes mechanism (fig. 26).

Many creation myths tell of an initial coming-to-be of the earth ordained by a benign force, a subsequent population that sinned or rebelled, and then purgation via a universal flood. The construction of Noah's ark certainly demanded mechanical skills, as the narrator in the third millennium B.C. *Epic of Gilgamesh* revealed:

> On the fifth day I laid the keel and the ribs, then I made fast the planking. The ground-space was one acre, each side of the deck measured one hundred and twenty cubits, making a square. I built six decks below, seven in all, I divided them into nine sections with bulkheads between. I drove in wedges where needed, I saw to the punt-poles, and laid in supplies. The carriers brought oil in baskets, I poured pitch into the furnace and asphalt and oil; more oil was consumed in caulking, and more again the master of the boat took into his stores. . . . On the seventh day the boat was complete. (*Gilgamesh*)

Carbon 14 isotopes have recently revealed the antiquity of mechanical and technological ingenuity throughout human civilization. In the early 1970s, a well was rediscovered in the Zhejiang province of China. The well was constructed over fifty-seven hundred years ago; it was 1.35 meters deep and 2 meters long, and flanked by over two hundred roundwood stakes bound with tendons and mortices (Shen 1985). Traces of windlasses (*lulu*), levers (*jiegao*), and pulleys used twenty-five hundred years ago can be found throughout the landscapes of the Shang Dynasty. The Book of Ezekiel (ca. 1000 B.C.) praised the Phoenicians for their skills in navigation and commerce and their grasp of strategic location. In Mesopotamia and Egypt, marvels of pottery, clay, and building materials continue to attract the archaeologist today. The fascination of making things work, of understanding how they work, as distinct from the delight in naming and claiming things, lies at the heart of mechanism as world-view. And for geography, it was the conquest of distance, particularly that of ocean and sea, that best challenged the mechanistic spirit (fig. 27).

The ebb and flow of enthusiasm for mechanism as root metaphor in Western intellectual life resonates well to the trilogy of symbolic figures Phoenix, Faust, and Narcissus. At the heart of those grand visions for West-

Fig. 26. Full-size model (3.6 m diameter) of a treadwheel from 1520, of the type that was used in Swedish industry during the eighteenth century. This replica was built at the National Museum of Science and Technology in 1980 by Svante Lindqvist and his students at the Royal Institute of Technology. (Photo courtesy of Svante Lindqvist.)

Fig. 27. Viking ship. (Sketch by Bertram Broberg.)

ern (post-Newtonian) science was a fundamentally emancipatory ideal. The seventeenth-century pioneers "sought all embracing frameworks, within which everything that exists could be shown to be systematically—i.e., logically or causally—interconnected, vast structures in which there should be no gaps left open for spontaneous, unattended developments, where everything that occurs should be, at least in principle, wholly explicable in terms of immutable general laws" (Berlin 1980, xxvi).

As in the Promethean tale, this quest has perennially led to ambivalent attitudes. While there has been the sense of triumphant mastery over fear and mystery in the universe, the disenchantment of the world has also led to feelings of loneliness and alienation (Lenoble 1969). An enduring motif in criticisms of modern science is that it robs nature of all mystery while denying humans their personhood and freedom. Even among scientists themselves, feelings of pyrrhic victory are sometimes expressed. For having "explained" how nature works, mankind now, in Monod's terms, lives as a gypsy "in a world which is indifferent to his music, just as indifferent to his hopes as it is to his sufferings and his crimes" (Monod 1972, 173). Yet, as Goethe sensed during the halcyon days of Enlightenment science, the Promethean urge would continue to elicit a Faustian dream, *eines Menschen Geist in Seinem Hohen Streben* (A mortal spirit in high endeavor).

Physis, Nous, and the Nature of Things

From the atomism of Pythagorean science, through the antiteleological fervor of the Epicureans, to the practical concerns of Roman engineers, sources of inspiration and cognitive foundations for a mechanist world-view can be found, many elements of which survive into the twentieth century. Already Anaximander (611–547 B.C.) objected to "single-element" explanations of the universe (*physis*) based on the primacy of "air," "heat," "matter" or, in the case of Thales (sixth century B.C.), "water" (Glacken 1967). For Anaximander, the universe was, to use a modern phrase, an "integrated system" governed by laws. *Nous* (mind or thought) was understood as a free-floating energy that influenced the diverse elements in their interactions, attractions, and repulsions. For Empedocles, the two primary forces were love and hate. To assume any teleological explanation of how the world worked, such as the existence of an extraterrestrial artisan, was, however, an unnecessary, even a repugnant hypothesis (Glacken 1967). The Pythagoreans, to whom the West owes, among other things, the Promethean myth, sought explanations of the terrestrial system via analogy with planetary movements, the "music and harmony of the spheres," whose vibrations suggested a mechanically ordered whole. What Prometheus stole from the gods were the mechanical arts and the secret of fire, enabling man to dominate the earth.

After the golden age of Socratic formism, the destruction of Periclean republicanism, and the faded dreams of Alexander the Great (356–23 B.C.), antiteleological views were again articulated. Theophrastus (369–285 B.C.), disciple of Aristotle, argued against the notion of final causes in history: so much seemed to be a matter of chance rather than necessity. On efficient causes one could debate, as later Greek and Roman scholars would do. Epicureans, doggedly objecting to Stoic teleology, strove to banish "gods" from the governing of the universe. Memories of the early atomists, such as Leucippus and Democritus (fifth century B.C.), would be recalled. As in many subsequent moments of social and political confusion, pleas for analytical rigor, "reductionism," and radical empiricism found an eager audience. Leucippus and Democritus envisioned a universe constituted by an infinity of components and particles of different shapes and substances, all constantly interacting within a universal void. The apparent diversity in the appearance of matter and form could eventually be explained via reductionist analyses of the ingredients and their functional interactions.

It was undoubtedly such brave beginnings that constituted the wellspring from which the classical world's most eloquent articulations of a mechanist world-view emerged—Cicero's *De natura deorum* and Lucretius's *De rerum natura.* These texts, rediscovered in late medieval times, proclaimed key doctrinal elements of a mechanist world-view which would claim perennial appeal:

1. The human mind can only grasp the nature of reality by penetrating details of its minute particles and ways in which these interact in diverse situations; the universe is ultimately constituted by the constantly changing associations of atoms in a void (Bailey 1926).
2. Teleological explanations of the universe are unscientific, unreasonable, and, at any rate, inscrutable. The earth, for example, could not conceivably have been designed as a fit abode for mankind: two-thirds of its surface being either too cold, too hot, too dry, or too wet for civilized human habitation. On the plenitude of nature, few observers could have doubts; however, in the conduct of life among species, there was always apparently a struggle for existence. From a human vantage point, for example, predatory beasts, poisonous plants, and a myriad of natural hazards suggested struggle rather than passive acceptance of "natural realities." If the gods had anything to do with this drama, Cicero and Lucretius both argued, they must be either malign or indifferent.
3. Whereas determinism and lawlike regularities could be discerned in the dynamics of physical forces ("inanimate matter"), these could not be found in the realms of living creatures. There one could discern necessities of survival, not lawlike determinisms, not divine fiat. Why should the gods care?
4. The great challenge for the human mind (*scientia*) was to eliminate fear and insecurity from terrestrial life: to enable humans to know and hence to control Planet Earth.

How other qualities of humanity were to be accommodated in this world-view still posed a puzzle for both Lucretius and Cicero. The latter stands out as a paragon of humanism in Western intellectual history, promoter of *paideia,* education in moral and intellectual virtues deemed requisite for the Roman citizen (see Chapter 2). While reductionist and deterministic in their views about the terrestrial system, both recognized that life itself possessed certain peculiarities:

> Again, if all movement is always interconnected, the new arising from the old in a determinate order—if the atoms never swerve so as to origi-

> nate some new movement that will snap the bonds of fate, the everlasting sequence of cause and effect—what is the source of the free will possessed by living things throughout the earth? (Lucretius, *De rerum natura* 2, 103)

The secret, Lucretius claimed, lay in mind (*intelligentia*), in intelligence which, like Anaximander's *nous,* transcended material determinism or externally imposed force:

> Thus you may see that the beginning of motion is made by the intelligence, and the action moves on first from the will of the mind, then to be passed onwards through the whole body and limbs. Nor is this like as when we move forwards impelled by a blow from the strength and mighty effort of another; for then it is clear that all the matter of the body moves and is hurried against our will, until the will has curbed it back through the limbs. . . . Wherefore you must admit that the same exists in the seeds also, that motions have some cause other than blows and weights, from which this power is born in us, since we see that *nothing can be produced from nothing*. For it is weight that prevents all things from being caused through blows by a sort of external force; but what keeps the mind itself from having necessity within it in all actions and from being as it were mastered and forced to endure and to suffer this, is the minute swerving of the first-beginnings at no fixed place and at no fixed time. (105)

Humanity, and especially human intelligence, the Romans assumed, was somehow exempt—morally and intellectually—from the deterministic laws of a mechanical universe. Furthermore, they claimed, humans bore a burden arising from their privileged Promethean power to rule and conquer the material world:

> We enjoy the fruits of the plains and of the mountains, the rivers and the lakes are ours, we sow corn, we plant trees, we fertilize the soil by irrigation, we confine the rivers and straighten or divert their courses. In fine, by means of our hands we essay to create, as it were, a second world within the world of nature. (Cicero, *De natura deorum* 2: 60, 151–52)

Technology, one of the first creations of humanity's artistic genius, celebrating freedom from extraterrestrial determinism (benign, evil, or indifferent), could thus become a universe unto itself, subject to lawlike determinism over which neither primeval nature nor supernatural whim should exercise control. Yet, in Roman as in several subsequent imperial contexts, the world-view of mechanism bore certain innate contradictions. The successful banishment of one teleology was often followed by the adoption of another. As with many a later empire in Western history, the Roman imperial system itself was to confront the test of managerial efficiency—in both its engineering and societal expertise—and eventually to crumble in the face of circumstances unanticipated in its elegantly integrated and thoroughly mechanistic world-view.

From Organic Whole to Mechanical Systems

The basically formistic conception of the universe, nurtured in medieval times, witnessed a virtual explosion during the Renaissance. The *aggiornamento* of spirit which attended the speculations of the Platonic academy and explorers of *terrae incognitae* in the fifteenth and early sixteenth centuries found inspiration in a basically organicist world-view (see chapters 2, 5). The surging wave of intellectual innovation no doubt hit many an intransigent cliff, for then, as at several later moments in Western intellectual history, two highly contrasting root metaphors, organicism and formism, confronted each other. Mechanism eventually offered a plausible alternative. Like organicism, it presented an integrated picture of reality; like formism, it stressed analytical rather than synthetic approaches. In late Renaissance times, mechanism again emerged, revitalized by a Faustian commitment to human progress. From a geographical vantage point, there are at least four major types of curiosity to which late Renaissance mechanism addressed itself, and whose fruits may still be recognized in twentieth-century science.

First, there was the fascination with the workings of things, with technology and engineering, epitomized in the ingenious inventions of Leonardo da Vinci (1452–1519) but also rooted in those occult crafts and magic that had for centuries defied the guardians of ecclesiastical orthodoxy. Of immediate relevance to geographical science were the challenges of navigation—compass and sail, harnessing the energies of water and wind; on land there were the prospects of canals and viaducts, castles and artillery. For the Renaissance pioneer, as indeed for Bacon and the metallurgical geniuses of northern Europe through the eighteenth century, *techne* connoted art, the practical arts, and also a radically new approach to knowledge (*scientia*). Insights into the nature of things based exclusively on mental processes or on evidence of the senses alone were equally suspect for Leonardo. As Vico would reassert two centuries later, men could only really understand what they themselves had created; only the artist, working with nature as engineer, craftsman, and experimenter, could find clues to nature's design. For Leonardo, the ideal analogy for the structure and mechanics of the entire universe was the human body itself, and this reveals his enduring attachment to an organicist rather than a mechanist world-view (fig. 33).

Mechanist ideas were also expressed in the realm of politics and power, epitomized in the work of Machiavelli (1469–1527), whose views were later echoed in those of Hobbes, Locke, and the fathers of the American Constitution. A statesman, satirist, and admirer of all things Roman, Machiavelli earned the title of devil from princes and potentates, for (as Voltaire said) he had actually "stolen their secrets." Angrily haranguing ecclesiastical and secular powers, he outlined a mechanical or, strictly speaking, a biophysical definition of society as a living system, which constructs a network of communication and commands so as to regulate the random uniformization or entropy (to use a modern phrase) which might otherwise result. As in Leonardo's system, the vital analogy was drawn from nature itself.

Thirdly, and most significantly, Renaissance times witnessed an expansion of horizons on cosmographic and astronomical puzzles, from the radically contextual beginnings of Nicolas of Cusa (1401–64) through the visionary hopes of Giordano Bruno (1548–1600), to Copernicus, Tycho Brahe, Galileo, and Kepler (de Santillana 1956; Koyré 1957). These astronomical discoveries eventually reverberated throughout all branches of knowledge, bringing radically new approaches to cartography and map projections. By the seventeenth century, a veritable revolution of consciousness had been achieved. Scholarly minds had moved from *theoria* to *praxis* (*poesis* to *ergon*), from contemplative to active knowledge, from filial obedience toward authority to self-propelled searching for scientific certainty. With this came a transposition of world-views from the synthetic ones of *organism* (e.g., Bruno and Leonardo) and *arena* (e.g., Nicolas of Cusa) to the analytically more incisive, albeit conceptually more restricted, approaches of *mechanism*. Koyré identified two essential features of this "mechanisation of world-view," which dominated seventeenth-century science after Descartes and Newton:

> The destruction of the cosmos and the geometrization of space, that is, the substitution for the conception of the world as a finite and well-ordered whole, in which the spatial structure embodied a hierarchy of perfection and value, that of an indefinite or even infinite universe no longer united by natural subordination but unified only by the identity of its ultimate and basic components and laws; and the replacement of the Aristotelian conception of space—a differentiated set of inner-worldly places—by that of Euclidean geometry—an essentially infinite and homogeneous extension—from now on considered as identical with the real space of the world. (viii)

While prospects of an infinite universe inspired by an organicist world-view delighted the philosophical imaginations of scholars such as Bruno, Thomas More, and Galileo, they frightened other pious souls more attuned to a view of the world as mosaic of proper forms. The Renaissance was essentially a time for *poesis,* and it is scarcely understandable in any other terms. Shakespeare (1564–1616) dramatized these characteristic ambivalences of the fifteenth century; in his plays, it is easy to discern a kinship of spirit with Dante, Machiavelli, and Leonardo. The Evil Angel, in Marlowe's *Doctor Faustus* (1604) counsels:

> Go forward, Faustus, in that famous art,
> Wherein all Nature's treasure is contain'd:
> Be thou on earth as Jove is in the sky,
> Lord and commander of these elements.

And Marlowe's Tamburlaine reflects:

> The thirst of reign and sweetness of a crown,
> That caus'd the eldest son of heavenly Ops
> To thrust his doting father from his chair
> And place himself in the empyreal heaven,
> Mov'd me to manage arms against thy state.
> What better precedent than mighty Jove?
> Nature that fram'd us of four elements,
> Warring within our breasts for regiment,
> Doth teach us all to have aspiring minds.
> Our souls, whose faculties can comprehend
> The wondrous architecture of the world
> And measure every wand'ring planet's course,
> Still climbing after knowledge infinite,
> And always moving as the restless spheres,
> Wills us to wear ourselves and never rest
> Until we reach the ripest fruit of all,
> That perfect bliss and sole felicity,
> The sweet fruition of an earthly crown.
> (Marlowe 1587, pt. 2, vii, 12–29)

Earthly rather than heavenly reward, experimentation and conquest rather than contemplation of mystery or beauty, the search for universal laws expressible in mathematical rather than ontological terms—these were the hallmarks of the Age of Reason spanning the period from Galileo (1564–1642) through Leibniz (1646–1716), whose major prophets also included Réné Descartes (1596–1650) and most significant of all for geography, Sir Isaac Newton (1642–1727).

Mechanization of the World Picture

The seventeenth and eighteenth centuries are frequently regarded as the take-off period for modern science. It was also a triumphant period for mechanism as root metaphor for reality as a whole with enduring reverberations in subsequent developments of medicine, engineering, and the philosophy of science. The logic was convincing indeed, and the picture of the world as machine was emotionally and aesthetically pleasing to minds now weary of Aristotelian categories, at times dubious about issues of faith and science, yet confident about the clarity and certainty of mathematics and physics. Mechanism, it could be argued, became a latent common denominator among scholars who took opposite sides on other dualisms whose modern expressions could also be traced to the same period: inductive versus deductive, empiricist versus rationalist, idealist versus materialist, and—most dramatic of all—the antinomy of "subjective" and "objective" truth so boldly propounded in the Cartesian distinction between mind (*res cogitans*) and matter (*res extensa*).

The overriding hope shared by seventeenth-century pioneers, as indeed by their Roman and Greek predecessors, was that nature could be understood as demonstrating precise mathematical laws; that such precise knowledge could free human minds from fear and superstition, thereby enabling humankind to master and transform the earth. They differed in method and orientation. Francis Bacon (1561–1662) and a long line of British empiricists—Mill, Locke, and Hume—favored experimental methods, inductive and empiricist approaches, and a concern about the applications of science to life and *techne*. According to Bacon, in the *Novum Organum,* "Human knowledge and human power meet in one; for where the cause is not known the effect cannot be produced. Nature to be commanded must be obeyed; and that which in contemplation is as the cause is in operation as the rule" (1901, Aph. 1, iii).

Réné Descartes, on the other hand, argued that truth was to be achieved by the exercise of pure reason armed with the deductive methods and precise language of mathematics. Evidence of the senses or imagination was suspect, but the mind through a process of systematic doubting could finally achieve certainty. His *Discourse on Method* (1637) wove a sophisticated case for regarding the human body as a machine understandable in the same precise categories through which one might understand stars and planetary bodies; even human emotions could be explained in terms of physics and hydraulics (J. Lynch 1985, 50–75).

With this profound change of metaphorical vision (Mills 1982), from the organicism of Renaissance Italy to the mechanism of France and England, which the Western intellectual world experienced in the seventeenth century, one also notes a shift from *poesis* to *logos* and a reopening of horizons toward *ergon*. Organic analogies between the human body and the dynamics of the earth, which inspired Leonardo's art, for example, became mechanical analogies on which to base scientifically analyzable methods for scholars of the eighteenth century. Mills (1982) offers a cryptic illustration here: while Renaissance scholars referred to the human body (anatomy, physiology, mind, and spirit) as an analogy for creation as a whole—the earth's hydrosphere analogous to the circulation of blood—the seventeenth-century mechanist used the alembic, or alchemist's, metaphor (fig. 28).

In the alembic distillation flask, water is heated to the boiling point, and steam so generated is then cooled to produce condensation in the head of the flask (Mills 1982, 246; Duncan 1951). Instead of a heart, the earth was now believed to have a central fire, heating incoming water flowing downward through subterranean passages from the oceans. Water then rises as steam to the earth's surface, where it condenses—for example, in mountains and hills. The alembic model could explain clouds, rain, thunder, and lightning, even earthquakes. This ushered in the era of James Watt (1736–1819), inventor of the steam engine. Harvey's treatise on the blood circulation system (1628) echoed many of Leonardo's ideas about water circulation.

Fig. 28. Triewald's glass tube with Cartesian divers. The human body is depicted as hovering between heaven and hell. Engraving in collection of lectures by Triewald in 1736. (Photo courtesy of Svante Lindqvist.)

Technical curiosity followed about how the theory could be tested—for example, Hales's pioneering measures of blood pressure (1733) and the functional role of air, temperature, mass, and gravity in the operation of the body machine (Lynch 1985). The "machines of marvellous efficacy" which Leonardo offered to the Duke of Milan in 1484, machines for war or peace, were only flights of a fertile imagination; the eighteenth century would turn them into realities (Bernal 1965; Schofield 1970).

Like their Roman forebears, enthusiasts for mechanistic explanations confronted the inevitable issues of humanness, free will, emotion, and determinism. Misgivings and soul-searching were also in the air. Pascal (1623–62), who shared much in common with Descartes on questions of faith and reason, was deeply troubled about the implications of a mechanistic world-view. Contemplating the implications of Galilean and Cartesian theories of the universe, he admitted, "The silence of infinite space frightens me." As to the idea of human emotions as subsumable under the rubric of body as machine, he declared: "The heart has its reasons that reason knows not of. . . . Do you love by reason?" (Pascal 1670). Ironically, this debate between Descartes and Pascal was conducted at a time when rationality was regarded as the essential defense against ecclesiastical dogmatism (witness Galileo's trial). Pascal's arguments against exclusively rational explanations were unwittingly interpreted as supporting irrationality and dogmatic blindness. Similar pathetic confusions over the meaning of "rationality" dotted the record of Western science and politics from the seventeenth century on.

Deus ex Machina

In Greek and Roman drama, a deity was frequently invoked who provided mechanical solutions to difficult situations. This deity became incarnate in seventeenth-century natural philosophy: Sir Isaac Newton's laws assumed an authority rivaling that of Moses' tablets. Alexander Pope proposed the following epitaph for the great prophet:

> Nature and Nature's Laws lay hid in night;
> God said, let Newton be! and all was light.

Newton's *Philosophiae Naturalis Principia Mathematica* (1687) marks the greatest milestone in the development of modern geographical science. In book 3 of the *Principia,* entitled *The System of the World,* the universal law of gravitation is presented. Mass, distance, and force were heretofore the essential categories through which the dynamics of the universe were to be understood. In Newton's view, Kepler's laws of planetary motion implied a central force, which varied inversely as the square of the distance between the sun and the planets. He postulated that the same force acted between any planet and its satellite, between the oceans, sun, and moon (e.g., in the tides), and, in fact, between the apple and the earth. In one compact mathematical formula, using the universal gravitational constant G, the gravitational force F between bodies with masses m_1 and m_2 at a distance r from each other could be accounted for:

$$F = Gm_1m_2/r^2$$

This notion of a "vital force," whose expressions could be stated mathematically, became a new religion among scientists and natural philosophers from the seventeenth century. Lavoisier in chemistry, Laplace in physics, Comte and later Durkheim in the social sciences, Buckle in history: scholars from a wide variety of fields grasped at this new key to understanding world reality.

As with other root metaphors, the appeal of mechanism cannot be entirely explained in epistemological terms. Its relevance for political iconography is graphically portrayed in the *Système du monde* (fig. 29):

The triumph of Newtonian cosmology in the seventeenth and eighteenth centuries is also understandable in contextual terms, as Halévy explained: "This was the beginning of the Utilitarian century, the century of the Industrial Revolution, of the economists and of the great inventors. The crisis had been brewing for fifty years: two names contemporaneous with the Revolution of 1688 symbolize the new era: 'Locke and Newton,' names which have become proverbially associated both in England and on the Continent" (1952, 5–6).

Réné Descartes had cautioned about the dualism of mind and matter, but the Newtonian theory could be applied to both. Why not probe the mechanisms of nerves and brains, emotions and perceptions? Why not indeed, wondered Hartley, whose *Observations on Man* (1749) provided a pioneering text for the behavioral psychology of succeeding centuries:

> These principles of Attraction and Repulsion of the several kinds, and of Vibrations, are dependent upon, and involved with each other, since this also is agreeable to the Tenor of Nature, as it is observed in the Body, in the Mind, in Science in general. . . . each part, Faculty, Principle, etc., when considered and pursued efficiently, seems to extend itself into the Boundaries of the others, and, as it were, to inclose and comprehend them all.
>
> And it would be no Objection . . . that we could not explain, in any definite Manner, how these things are effected . . . nor would this be to reason in a Circle, more than we argue, that the Heart and Brain, or the Body and the Mind, depend upon each other for their Functions; which

1638

DISPOSITION DES PLANETES AU MOMENT DE LA NAISSANCE DU DAUPHIN ·

Fig. 29. *Système du monde*. This seventeenth-century engraving depicts the configuration of the world system "at the moment of the birth of Louis the Great," illustrating the political significance of royal power within a cosmic order ruled by Copernican astronomy and Cartesian physics. (Reproduced by permission of the Bibliothèque Nationale, Paris.)

> are undeniable Truths, however unable we may be to give a full and ultimate Explanation of them. (110)

The legacy of seventeenth- and eighteenth-century mechanism in psychology and other social sciences was to become a virtual Trojan horse for subsequent generations, not least for the behaviorists of the mid-twentieth century (Haldane 1939). It was in the languages of Cartesian and Newtonian science that the vision of scientific humanists such as D'Holbach, DeLamettrie, and the *Encyclopédistes* were articulated (Diderot and d'Alembert 1751–80; Diderot 1818; Dainville 1941). "We are on the verge of a great revolution in the sciences," Diderot write in 1753. "To judge by the inclination that our writers have toward morals, fiction, natural history, and experimental physics, I feel almost certain that before 100 years are up, one will not count three great geometers in Europe" (Diderot 1818, 1:420). The triumph of mechanism over formism was echoed in the scientific scenarios for future society advanced by Auguste Comte (1789–1857) and Emile Durkheim (1858–1917). World as mosaic was engulfed in the revolutionary idea of world as mechanical system. Mechanism attracted rationalist and empiricist, theist and atheist, idealist and materialist: it was a world-view espoused not only for its intellectual appeal but also for its salience within the sociopolitical contexts of eighteenth-century Europe (Foucault 1975). Not only could mankind now be in charge of creation,[1] but the state could be regarded as the perfect machine to monitor all aspects of life and natural resources.

Utilitarianism and the Industrial Revolution may have characterized the European contexts in which mechanism found welcome as world-view. Across the Atlantic, mechanism would also win appeal in the curricula of New England's colonial colleges from the mid seventeenth century on (Warntz 1964). A study course on The Uses of the Globes was among the most popular: curiosity was evoked about mathematical regularities in global dynamics, the sphericity of the earth, magnetic fields, and analogues between terrestrial and planetary motion. Newtonian geography presented an image of the world which could be shared with colleagues in history, politics, and law; it was also a useful field of study for anyone wishing to ply the Yankee Clipper sails, prospect in trade or commerce, or pioneer on the westward-moving frontier of a new land. After the War of Independence, delegates met in Philadelphia to frame a new Constitution (1787), many of whom were graduates of these colonial colleges and found a common denominator of interest in Newtonian ideas of natural law. Positive law should follow the dictates of natural law: this new society should be regulated in terms that reflected the nature (constitution) of that society. They eagerly affirmed the integrity of parts in the whole: "'With genuine enthusiasm,' wrote Woodrow Wilson in 1908, 'albeit without approval, the founding fathers followed the lead of Montesquieu under whose hand politics turned into mechanics. The imitation of the checks and balances of the solar system was a conscious one, yielding a balancing of powers among president, congress, and the courts, and in the congress between big and little states'" (Warntz 1964).

As Arthur Lovejoy noted, the American Constitution embodied two distinct but contemporaneous ideas about humanity and world. Alexander Pope had voiced a view of mankind's "warring passions," and the conviction that by bouncing evil things against each other something good could eventually emerge (Lovejoy 1961). It took only a short flight of metaphorical imagination to combine this idea with the Newtonian theory of mechanics.[2]

Warntz claimed that eighteenth-century political science, based as it was on mechanics, was far more sophisticated than the physical science of the day (Warntz 1964). Various forms of physical energy and their conversion, he claimed, were yet to be fully described, but the analogous social energies such as reason, feeling, and authority and their rich interplay were spelled out in the judicial, legislative, and executive branches of the Constitution.

As for geography's claims of relevance to the more pragmatic concerns of a pioneering society, Warntz found a less reassuring record. Although eloquent support for applied geography was voiced by Jared Sparks (class of 1815), who later became president of Harvard University, the subject was dropped from the curriculum only one year later. As with several other cycles of geography at Harvard, contextual conditions played a major role (Buttimer 1985b). Warntz's account of the Harvard story suggests, among other things, that among geography's claims for status as an academic field, mechanism superseded mosaic in cognitive appeal, and *logos* rivaled *ergon* in vocational appeal.

In European settings, these tensions would take altogether different forms, following political and ideological currents then salient in burgeoning nations and empires. Yet apart from the strictly mathematical and cartographic aspects of geographic inquiry, neither mechanism nor mosaic could possibly compete with the other global world-view that was in the late eighteenth-century air. From the early 1800s on, organism reemerged, and tensions between romanticism and Enlightenment reached dramatic heights on issues of environment and culture. Modern academic geography was launched in North America with an organicist metaphorical vision (see Chapter 5), and its main alternative was that of mosaic. Might it not have been conceivable that mechanism would have provided a compelling alternative, as it had in Renaissance times? Many late nineteenth-century pioneers of sociology launched their disciplinary programs with organic analogies between physical, biotic, and cultural phenomena. A second generation would strive to implement and operationalize these ideas, often reverting to mechanistic frameworks. In fact, Prigogine and Stengers claimed that during the nineteenth century

> the term *Newtonian* was now applied to everything that dealt with a system of laws, with equilibrium, or even to all situations in which natural order on one side and moral, social, and political order on the other could be expressed in terms of an all-embracing harmony. Romantic philosophers even discovered in the Newtonian universe an enchanted world animated by natural forces. More "orthodox" physicists saw in it a mechanical world governed by mathematics. For the positivists it meant the success of a procedure, a recipe to be identified with the very definition of science. (Prigogine and Stengers 1984, 29)

Among geographers, however, there were mostly lukewarm and even occasionally hostile attitudes expressed about mechanism as root metaphor for reality. To shed light on this, one needs to look at the root metaphor itself and its *causal adjustment theory of truth,* as described by S. C. Pepper (1942, 186–279).[3]

A Causal Adjustment Theory of Truth

Levers and pulleys afford the common-sense grounding for mechanism as world hypothesis and its root metaphor, "the machine." Pepper offers the example of a lever consisting of a bar resting on a fulcrum. The balance of forces could be described in terms of:

$$W_A \times L_A = W_B \times L_B$$

where W_A is the weight of a mass placed at a distance L_A from the fulcrum and W_B the weight of a mass on the other side and at a distance L_B from the fulcrum.

A distinction is drawn between "discrete" and "consolidated" mechanism. Discrete mechanism, as illustrated in the atomic materialism of Lucretius, is based on the assumption that many of nature's structural features are loosely related. Space is distinct from time, primary qualities (weight, pressure, quantity) are distinct from the field of locations, and natural laws are distinct from each other. Natural laws are discrete and separable from the masses on which they operate. In consolidated mechanism, on the other hand, such distinctions cannot be claimed. Instead of the discrete particle in a spatiotemporal path, or instead of the discrete laws of mechanics, a cosmic geometry (or geography) is envisioned to describe the spatiotemporal whole (212). The chief catalyst for consolidated mechanism has come from the general theory of relativity in which mass is interpreted in terms of a gravitational field, that is, a spatiotemporal-gravitational field, not dissectible into discrete categories of location, time, and gravitation. In Pepper's view, the machine might be the appropriate root metaphor for discrete mechanism, and the electromagnetic field might best capture the world-view of consolidated mechanism, one resembling a kind of gelatin or crystallized geometry with an intricately woven internal structure (186–279).

There is tension between these two types of mechanism: ultimately discrete mechanism tends to merge with formism, whereas consolidated mechanism tends to merge with contextualism. Common categories of analysis are suggested by the mechanistic approach to reality.

Primary categories include:

1. Locations in time and space which determine the mode in which the machine functions;
2. Primary qualities, e.g., size, shape, motion, weight, or mass number;
3. Mathematical relationships or laws which hold among parts of the machine.

Secondary categories include:

4. Qualitative features such as color, texture, smell, etc., which do not in themselves directly influence the functioning of the machine;
5. Principles whereby these qualitative features are combined with the primary categories (1–3);
6. Secondary laws governing relationships among secondary categories.

Mechanism as world-view has held appeal for both materialists and idealists. In Pepper's definition, a materialist such as Hobbes would focus primarily on primary categories, ignoring or denying secondary ones, whereas an idealist such as Berkeley would focus primarily on the secondary categories. Neither position, in Pepper's view, is cognitively credible. The two sets need each other; after all, the cognitive evidence for the structure and details of the cosmic machine described through the primary categories comes entirely from materials within the secondary categories.

Traditional discrete mechanism was founded on a theory of elementary particles distributed in space and time (Reichenbach [1951] 1973). Primary qualities such as size, shape, motion, solidity, mass, and number were regarded as inhering in spatial volumes and as continuing unaltered through time except for changes of position in the field according to basic physical laws. Space and time were fundamental, and a Newtonian conception of an infinite three-dimensional manifold afforded an adequate system of possible places. To this could be added an absolute time manifold of externally related dates, and later a fourth dimension of the field of locations. The existential particular could be regarded as a space-time particular: all locations (here-and-nows) had a determinate field structure. Laplace's dream in *Système du monde* was firmly anchored in the possibility of a completely deterministic universe, explainable in terms of Newtonian laws of motion (Laplace [1796] 1966). Given the knowledge of the configuration of masses in the spatial field at any one time and the laws that operate upon these masses, then all possible configurations, past or present, could be predicted.[4]

Quantum mechanics introduced probabilistic concepts and demonstrated the intrinsic uncertainty of simultaneous measurements of position and velocity of a particle. The vision of Laplace's demon, namely, to determine these qualities exactly for all particles in the universe, would become impossible, both practically and theoretically. Late twentieth-century theorists argue that the information present at any given time would be insufficient to predict the future course of the universe (Hawking 1988).[5]

The most disconcerting puzzle for mechanists ever since Descartes is the "mind-body" problem. Human cognition depends on secondary qualities—sensory and other qualities of experience which are not reducible to the primary qualities of size, shape, motion, mass, and number. Mechanisms imputed to the cosmic machine are not directly visible, and ordinary perception does not fit with the spatiotemporal structures of the cosmic field. One experiences sights, smells, feelings, and mental states. Several theories have attempted to explain connections between the so-called secondary qualities and the primary qualities of the world machine. How might one explain those sensory qualities of human perception, which seem so superfluous to the great machine?

This incongruity has led to extremes of subjective idealism, for example, propositions that only mind or consciousness exists. One is aware only of one's own organism, and hence draws inferences from this to the behavior of animals or plants. Complex mental states might be examined via analogies from chemistry: laws of association in mental states could be seen operating upon elements to produce more complex mental states. Could these laws of association be regarded as implying introspective manifestations of psychological laws, which in turn could be regarded as complex operations of mechanical laws? There was something of this implicit credo in Locke's theory of cognition, but twentieth-century psychologists have repeatedly questioned such a solution. Gestalt psychologists refused to reduce thought to mental elements and maintained that the whole needed to be understood before parts could be appropriately analyzed. Lewin's "field theory" afforded at least one more satisfactory account of perception (Lewin 1952).

Mechanism, in all its expressions, has had difficulty in coping with the puzzle of relating mind and matter while still maintaining its causal adjustment theory of truth.[6] There were other objections on moral, aesthetic, and cognitive grounds, which were raised by the late nineteenth-century founders of geography as a discipline. A closer look at the contexts of potential professional practice, the environmental interests of potential sponsors and audience, and the burgeoning of new analytical movements within science itself during that critical foundational period shed light on the lukewarm or hostile attitudes displayed by geographers toward the root metaphor of mechanism.

Geography and "Systems"

The predominant metaphorical styles of the late nineteenth and early twentieth centuries, in Europe as well as in North America, were those of organism and mosaic. Intellectual impasses over the definition of natural regions and their boundaries, as well as over relationships between human and physical branches of the field, could ultimately be traced to fundamental differences between these two world-views (Buttimer 1985b). Why, then, one wonders, did mechanism not appear as a reasonable alternative?

There were scattered examples of mechanistic thought in the early twentieth century. Mark Jefferson waxed eloquently on the "civilizing rails"; Jean Brunhes explored the irrigation systems of Iberia; and Vidal de la Blache wrote on cities as *axes de crystallisation* shaped by circulation routes. Few questioned the applicability of deterministic laws to the study of natural systems, but most shrank from the prospect of imposing such frameworks on study of human habitats or the morphology of landscape. From its very beginnings as an academic discipline, geography had enjoyed intimate ties with geology and botany, fields in which tensions of mosaic and organism were already dramatic, and also with history and anthropology, fields in which tensions between organism and arena were already assuming keen political significance. William Morris Davis, passionately committed to evolutionary modes of interpreting landscape morphogenesis, won applause in North America, far exceeding that given to Grove Karl Gilbert, whose cognitive style bore closer resemblance to that of classical mechanism (D. Sack 1991).[7] A number of irreconcilable debates occurred within these neighboring disciplines during the nineteenth century, such as those related to the antiquity of the earth, the evolution of species, and especially the relationships between human civilizations and milieus. Competing claims were often laid at the judgment bench of biblical interpretation rather than that of experimental science.[8] Irrationality in some of these debates might indeed have encouraged geographers to turn toward mechanistic modes of explanation.

In terms of day-to-day practice, however, one wonders if they had time to think critically about the epistemological status of their research. Many of the founders of geography as an academic discipline held positions in one-person departments with minimal technical infrastructure and personnel. *Poesis* might be directed toward persuading academic colleagues of the nature and value of the field as well as convincing sponsors about the value of exploration and the world map. The bulk of one's energies, however, were channeled toward *paideia:* teaching, leading field excursions, and preparing manuals and text materials. Organism offered appropriate metaphorical grounding for the poetic side, and mosaic certainly offered excellent grounding for the educational one. The practical relevance of geography for society (*ergon*) would remain at best an implicit goal for these overworked pioneers; besides, they were eager to present geography as a scholarly field rather than just as a practical art. It would be in terms of *logos* and the research frontiers of the discipline that mechanism staked its cognitive claims. Already around the turn of the century, its analytical potential had been strongly bolstered by the growing appeal of positivism among European scientists.

The lukewarm and occasionally hostile attitudes of leading geographers toward mechanism might have reflected a skepticism about positive science as much as doubts about the human implications of such a world-view. For one thing, embracing positivist analytical methods involved choices not easily reconcilable with another priority at the time, namely, establishing a special identity for the field among the humanities and natural sciences. Geographers sought a core of theory and method which could embrace both human and physical phenomena in an integrated fashion. There were scholars of positivist commitment on both sides who argued vehemently for a separation of "physiography" and "ethnography," the one answerable to the scientific protocol of geological science, the other to that of emerging social science, but the main corpus of leading geographers sought to maintain unity within the discipline. They argued that it was precisely this ability to elucidate interactions between humanity and milieus which gave geography its distinctive character among academic fields. The cardinal presupposition underlying positivist method was *ceteris paribus* (other things being equal). Until such assumptions could be ventured, mechanism in its positivist analytical expressions could not gain an audience among geographers. The integrative character of its promised results would not, at least initially, imply an integration of human and physical orders of reality which would do justice to both. Rather, it would virtually force a cleavage between them. The first flowerings of twentieth-century mechanism in human geography were in the analysis of demographic, economic, transport, and urban phenomena—just those fields in which it could be somehow argued that the natural milieu could be regarded as either irrelevant or so technologically transformed that the cherished assumptions of *ceteris paribus* could be upheld. In physical geography, mechanism found a welcome among those who studied specific layers of the natural environment—water, climate, soils, minerals, or other domains where laboratory testing or otherwise controllable experiments were conceivable.

Pioneers of the discipline of geography had academic backgrounds in liberal arts, such as history or anthropology, and in the sciences, such as geology, botany, or meteorology. In most of these fields, there had been some skepticism about mechanistic interpretations of reality. On the side of the natural sciences, there were tensions between founders who had drawn inspiration from natural history, botany, and biology and those with more specialized training in mineralogy, lithospheric processes, or atmospheric processes. Perspectives on nature, space, and time differed markedly at both ends of this antinomy. Cartesian (and eventually Kantian and Newtonian) science projected a vision of the world in the categories of Euclidean geometry, implicitly assuming absolute (Platonic) notions of time and space as a priori categories of human understanding. In understanding the nature and dynamics of geological processes, time could be regarded in mechanistic terms as a dimension of reality potentially measurable in isometric units within an infinitely expandable universe, or alternatively as a clock ticking within observable machines already scrutinizable via experimental methods (Adams 1968). "When we trace the parts of which this terrestrial system is composed, and when we view the general connection of these several parts, the whole presents a machine of a peculiar construction by which it is adapted to a certain end" (Hutton 1795, 1: 3).

Such an approach could scarcely serve to elucidate the course of human history on earth. The history of Planet Earth within the cosmos was one kind of challenge, but the history of humanity was quite another. Was there an "arrow" through time in human history—an inevitable march from barbarism to civilization, or from Yahweh's self-revelation to the Christian *parousia*—or should terrestrial experience be more appropriately construed in terms of cycles, in its seasonal and diurnal manifestations, as nature suggested? Tensions between the "wisdom of God" and the "deus ex machina" were eventually to reintroduce issues of temporality in eighteenth-century discourse about the universe. Hutton's *Theory of the Earth* (1795) and Lyell's *Principles of Geology* (1830), the foundational texts for physical geography as an academic discipline, reveal the tensions between "arrow" and "cycle" in the interpretations of time in terrestrial history (S. J. Gould 1987).

On the "human" side, some of the geographer's closest associates (historians and anthropologists) showed little enthusiasm for mechanistic models of society, frequently using "geographical factors" as a defense for their antideterministic arguments (Febvre 1925). The geographer's own search for explanatory laws in the formation of landscape was a cautious one. In the early days of *la géographie humaine,* Vidal de la Blache acknowledged a number of causal chains (*enchaînements*) in the natural world, but warned that virtually all events in human history were *"frappé de contingence."* Lukermann summarized the French School's attitudes toward contingency:

> Briefly it is because any part of a whole is "dependent on" and has value only in terms of, all other parts of the whole; that is, theoretically, every event in the world is conditioned by every other event in the world, past and present. Practically, then, any event is the result of a number of causal chains intersecting at some point in time-space. The event is caused, but nevertheless not determined; rather it is contingent on the indispensable but fortuitous intersection in time-space of multiple chains of causation. (Lukermann 1965, 130)

It would be difficult to understand the interplay of metaphor and vocational meanings during this foundational period without reference to the milieus in which the discipline first found its anchoring. The teaching and research goals of European geographers and historians at the turn of the century were intimately linked with those of their major sponsors, namely, nation-states and empires. It was in an era of national self-confidence and often expansionary politics that pioneering steps were taken in the production of text materials and research exemplars, especially during the period before World War I. While a great Pax endures, to use Umberto Eco's phrase, scholarly energies may focus on the status quo; formism offered an infinite variety of specific tasks, from surveys of home areas to atlases of world geography. Within a formistic framework, geographers could address public interest in order and inventory, as in studies of *Heimaten* and *pays.* Public interest in local and regional identity could also be served thereby, but horizons would be local. Organism would serve the interests of linking local to global, supplying a romanticist perspective on landscape and history, especially in settings where one's own capital city could be regarded as the center of the universe (see Chapter 5).

Mechanism has characteristically been the favored metaphor of reformers critical of the status quo or eager to explore its structural dynamics (H. White 1973). Not until the evening of the great empires, namely, World War I for many, did mechanism find favorable circumstances in which to play an emancipatory role.[9] Intellectually speaking, the ground had already been prepared by "radical" advocates of positivism. Janik and Toulmin (1973) suggest that there was something of a Renaissance esprit among those early twentieth-century members of the Vienna Circle. Specialists from fields as diverse as music (Schoenberg), psychology (Freud and Jung), physics (Mach), linguistics (Wittgenstein), and biology (Lorenz) found common denominators of goal and analytical method as together they undertook an autopsy of the Hapsburg Empire and its pyramid of power. With mechanism as root metaphor and positivism as analytical guide, they explored the limits of Enlightenment reason, of language and power, of acoustics and aesthetics, of the human psyche and the nature of space and time. Leonardo-style, they toyed with the mechanics of sensory perceptions, phonetics, and politics, sharing a common ethical commitment to Comtean ideals. Their creed would spread, and some of the pioneers themselves would migrate across the Atlantic before midcentury.

If World War I catalyzed contextual reflection and the search for alternative world-views among Europeans, the 1929 stock market crash halted the Roaring Twenties in America. Though the majority of human geographers were content with chorographic and formistic inventory and regional description, there were a few—at Chicago, for example—who again entertained organicist ideas in seeking solutions to the country's environmental problems and new approaches to the management of natural resources (White, G65). In the late thirties, "the lights were on all week-end on the Potomac," Gilbert White remembers; "the time was ripe for new approaches." Organism soon yielded ground to mechanism, *poesis* was channeled toward *ergon*, and problem solving and management of environmental resources became a political priority, at least for a few.

In urban and transportation research, the interwar period also witnessed an exploration of Newtonian ideas. R. M. Haig's (1928) multivolume study, *New York and Its Environs,* encapsulated much of the mechanist-cum-positivist élan of the early twentieth century. Echoing the Enlightenment (*Wealth of Nations*) dream and Pope's faith in a positive outcome from mankind's "warring passions," this monumental research enterprise made a profound impression. It was welcomed in Stockholm, the city that staged a thoroughly functionalist display of *Bauhaus Bewegung* architectural style at its 1930 World Exposition. Hans W:son Ahlmann decided that a similar study should be done on Sweden's capital city (William-Olsson 1937; PG, 153–66).

Functionalist perspectives on human habitat and circulation patterns were also gradually developed during the twenties and thirties in continental Europe. Vidal de la Blache's *France de l'Est* (1917) and Albert Demangeon's *Les îles brittaniques* (1927a) pointed toward the need for more dynamic approaches to regional geography; studies of ports and commercial geography, as conducted at Hamburg, Bordeaux, and Stockholm, essayed mechanistic approaches to space and interactions between places (Demangeon 1927b; E. Kant 1934a; Hannerberg 1957). The cardinal role of circulation as potentially disruptive force and eventually as facilitator of innovation was a theme that ultimately undermined the hegemony of formal, chorographic approaches to urban geography. Bobek's *Innsbruck* (1928) struck a new chord in the perennial discussions of site versus situational forces in the shaping of urban life and landscapes (see also Lavedan 1936; Capot-Rey 1946).

Mechanism in its interwar expressions appeared to sound an emancipatory note: functional perspectives were regarded as liberation from the prison of inherited forms. Such, at least, is the impression yielded from a rereading of Kant's work in Estonia (figs. 30, 31) and of his visions for integral development of a Balto-Skandian region (E. Kant 1935, 1946).

At the Stockholm School of Economics, meanwhile, Bertil Ohlin and Gunnar Myrdal were advancing new economic theories that later influenced Roosevelt's welfare policies. At Harvard, visiting German theoreticians such as August Lösch and Alfred Weber shared ideas with American scholars such as John Kenneth Galbraith and Milton Freedman about the integration of spatial factors in conceptions of economic life. C. C. Colby's essay, "Centrifugal and Centripetal Forces in Urban Geography" (1933), was certainly a milestone for the audience in the United States; Chauncy D. Harris and Edward E. Ullman's "The Nature of Cities" (1945) became the pacesetter for at least two decades of urban geography in that land.[10]

Wartime service by trained geographers on both sides of the Maginot Line may indeed have played an important role in this transformation of metaphorical vision from form to function, pattern to process, as well as in the shift in research concern from areal differentiation to spatial interaction (PG, 185–96). "World War II was the best thing that happened to geography since the birth of Ptolemy," Kirk Stone reminisced (1979, 89). Substitute "mechanism" for "geography" in that phrase, and in historical perspective few could disagree. In the heady optimism of postwar reconstruction, aid from the Marshall Plan, and regional development projects, on the European side, and continued U.S. commitment to science, technology, and economic growth, systems thinking became the hallmark of serious research in the fifties. A missionary band proclaimed a "New Geography," where notions of place and space as container of objects and patterns were replaced

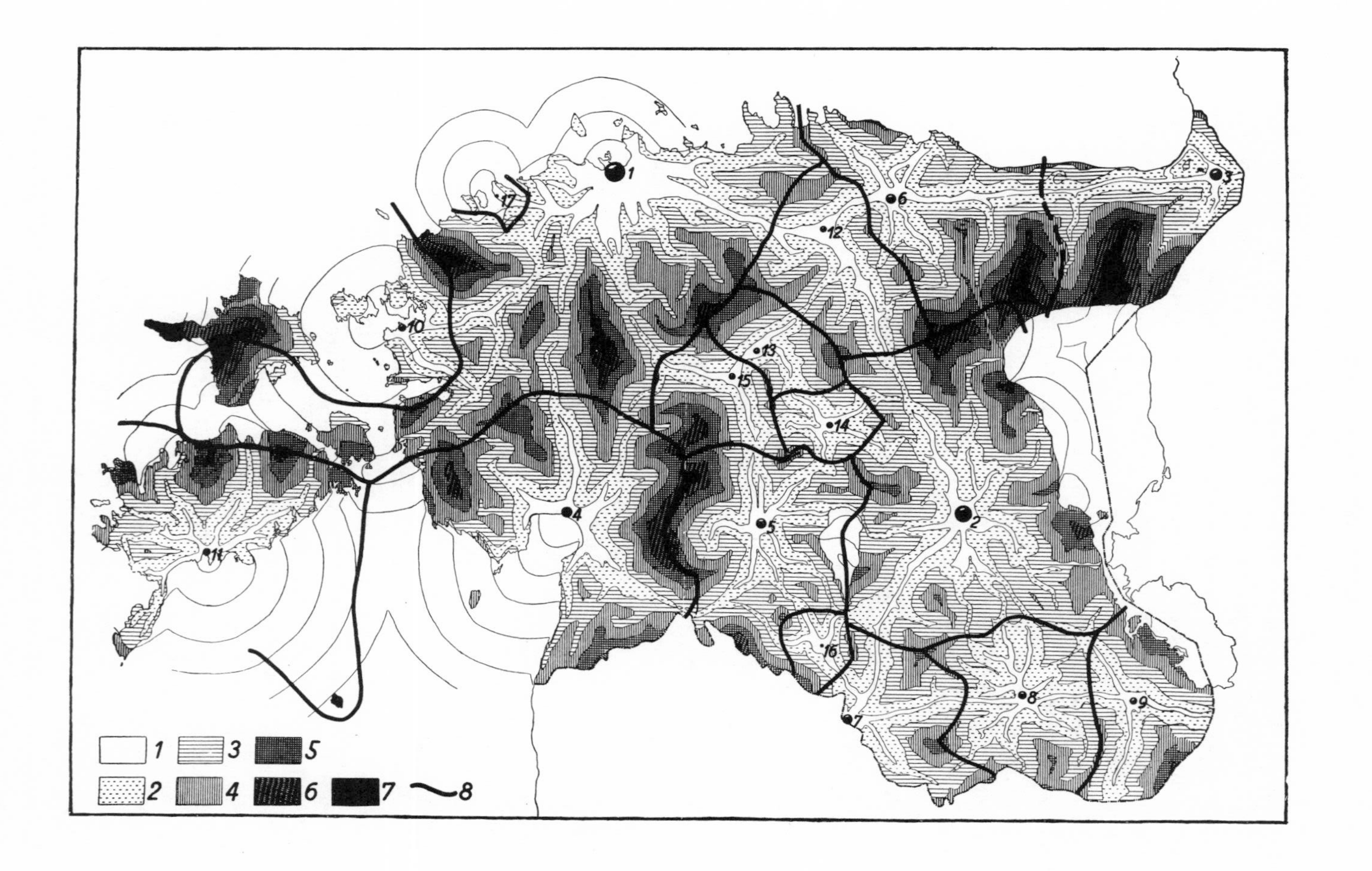
1
3
5
2
4
6
7
8

by notions of space as topological surface, nodes, and networks (Ullman 1954; Alexandersson 1956; Ajo 1953; Ackerman 1963).

The gravity model went a long way in explaining distances between places. The volume of interaction between two cities could be modeled by multiplying their populations (surrogate for mass) and thus achieving a measure of all potential flows between them, and by dividing by some measure of "distance friction" between them:

$$I_{ij} = kP_iP_j / D_{ij}^b$$

where

I_{ij} = the interaction between place *i* and place *j*

P_i = the population of place *i*

P_j = the population of place *j*

D_{ij} = the distance between places *i* and *j*

k = constant

b = exponent

At least with respect to commodity flows between cities, Ullman refined this via three key considerations: complementarity, transferability, and intervening opportunity (Ullman 1940–41). The Rank-Size Rule offered hypotheses regarding the relative sizes of cities within any national system: other rules relating to economic base, comparative advantage, and functional complexity demonstrated the intimate relationships between circulation and habitat (Berry 1964, 1972). Behavioral approaches allowed graphic portrayals of cultural influences on the choice of services available at urban centers (fig. 32).

Eventually geographers sought models of consumer behavior, "decision-making" and "choice," underlying all the idiosyncrasies of individual perceptions, needs, or values (fig. 33).

As an integrated world-view, mechanism offered unlimited scope for both speculative (*poesis*) and practical (*ergon*) aims of postwar urban and economic geography (Usard et al. 1960). Its appeal, however, was generally made in terms of cognitive power (*logos*). This was probably felt first among

Fig. 30. Fields of urban influence in Estonia: Polycentric isochrones of time-distance to Estonian cities. The numbers indicate urban settlements: 1, Greater Tallinn Tallinn-Nömme; 2, Tartu; 3, Narva; 4, Pärnu; 5, Viljandi; 6, Rakvere; 7, Valga; 8, Vöru; 9, Petseri; 10, Haapsalu; 11, Kuresaare; 12, Tapa; 13, Paide; 14, Pöltsamaa; 15, Türi; 16, Törva; 17, Paldiski. The heavy curved lines (8) indicate boundaries between urban hinterlands; the shaded zones (1–7) symbolize time-distance intervals from 1 to 6 hours and over 6 hours. (E. Kant 1934, kartenbeilage x. Courtesy of Geography Department, University of Lund.)

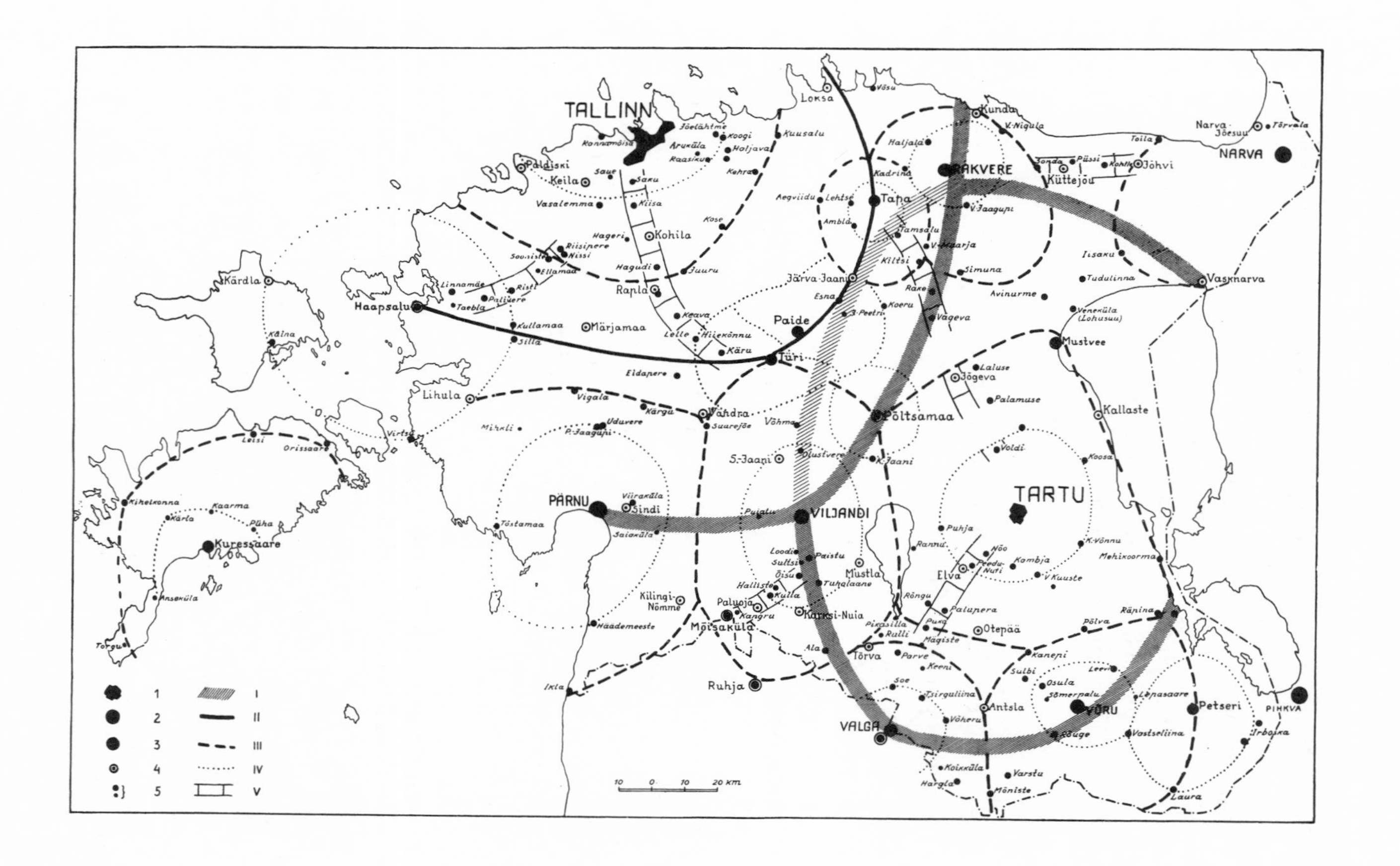
TALLINN
NARVA
Narva-Jõesuu
Loksa
Kunda
RAKVERE
Küttejõu
Jõhvi
Vasknarva
Tapa
Järva-Jaani
Paide
Türi
Rapla
Kohila
Keila
Paldiski
Haapsalu
Märjamaa
Kärdla
Lihula
Vändra
Põltsamaa
Jõgeva
Mustvee
Kallaste
TARTU
PÄRNU
Sindi
VILJANDI
S.-Jaani
Mustla
Elva
Otepää
Tõrva
Karksi-Nuia
Mõisaküla
Paluoja
Kilingi-Nõmme
Ruhja
VALGA
Antsla
VÕRU
Petseri
PIHKVA
Kuressaare
1
2
3
4
5
I
II
III
IV
V
10 0 10 20 km

physical geographers; for example, in Birot's geomorphology and in Strahler's dynamic concepts of hydrological systems and erosional processes, to be pursued by "process geomorphologists" during subsequent decades (Birot 1959; Strahler 1952). The Lund School pioneered in models of migration and diffusion and in fields of urban influence; the Stockholm Congress (1960) marked a high point in the success of mechanism among human geographers internationally (E. Kant 1946; Hägerstrand 1953). Ackerman's presidential address to the AAG in 1962 left little doubt that the frontier lay on the horizons of systems analysis (Ackerman 1963). Berry (1964) wrote that "geography's integrating concepts and processes concern the world wide ecosystem of which man is the dominant part."

The appeal of mechanism for *ergon* was by no means less attractive (Perroux 1954; Lösch 1954; EFTA 1974; Pred 1977). J. W. Forrester wrote in the late sixties: "An urban area is a system of interacting industries, housing, and people"; and, "The urban system is a complex network of positive- and negative-feedback loops. Equilibrium is a condition wherein growth in the positive loops has been arrested" (1969, 1, 121). "Dynamic systems" were voguish words in engineering, biological, social, psychological, and ecological research throughout the sixties.

During the sixties, attempts were being made to move beyond the determinism involved in Newtonian models, and statistical mechanics sought to link spatial interaction models and information theory, for example, with measures of entropy maximization (Wilson 1970; W.K.D. Davies 1972). In practice, a decisive reorientation was occurring in academic geography toward *logos* and *ergon,* as research in the field became more dependent on external funding. Its *paideia* and *poesis* were redirected toward these apparently more powerful ways of conceptualizing, managing, and planning the human use of the earth (Isard et al. 1960; Lebret 1961). The overwhelming enthusiasm for mechanist modes of interpreting human behavior in space (Kates and Wohlwill 1966), which was demonstrated in Euro-America throughout the late fifties and early sixties, no doubt reflected general public opinion, growing economic prosperity, and visible miracles of technology

Fig. 31. Fields of urban influence in Estonia: Systems of central places. Numbers 1–5 indicate rank-order size categories: 1, cities; 2, small cities; 3, rural towns; 4, agglomerated settlements; 5, hamlets [*Hakelwerke*]. Numbers I–V indicate Hinterland zonation around central places: I, Zone of city service reach; II/III, Outer and intermediate belts of reach for urban services; IV, Belt of reach from lower-order centers; V, Radial order of central places: Break of transportation points in the radial ordering of service centers. (E. Kant 1934, kartenbeilage xi. Courtesy of Geography Department, University of Lund.)

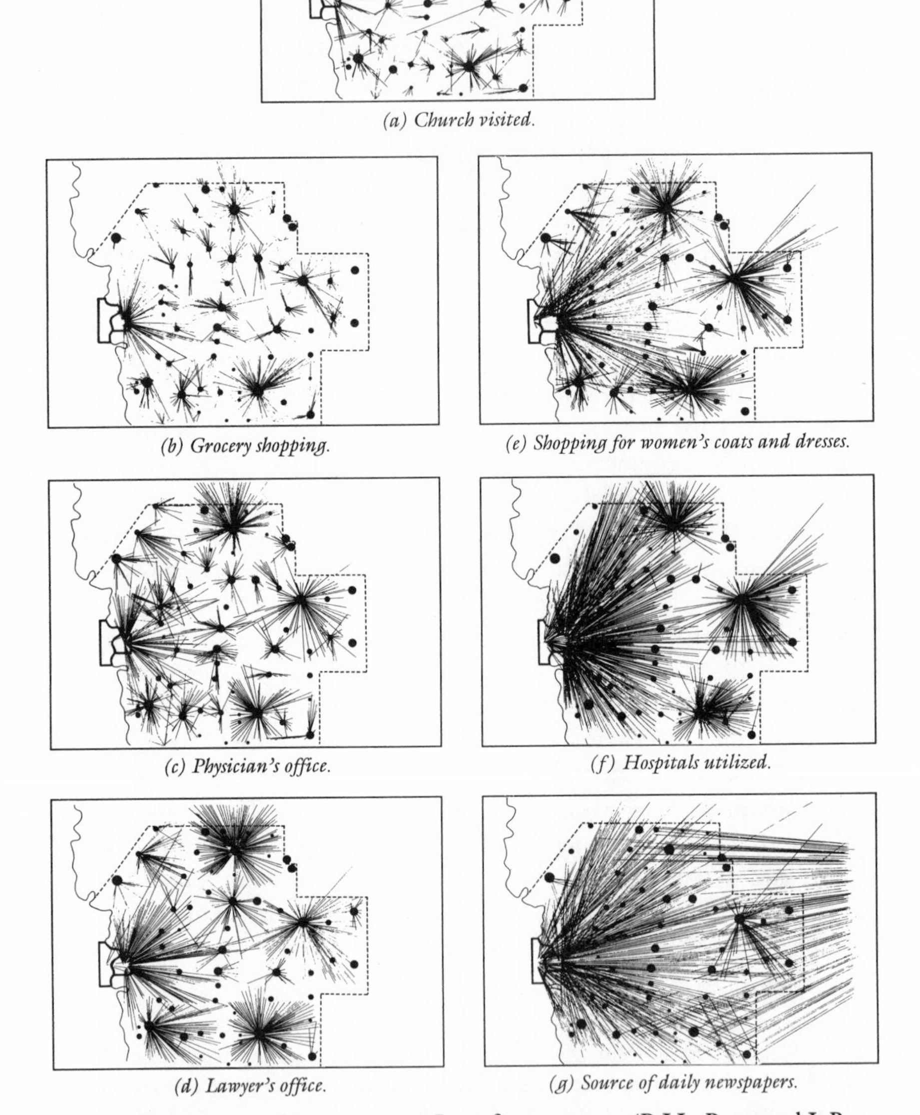

(a) Church visited.

(b) Grocery shopping.

(e) Shopping for women's coats and dresses.

(c) Physician's office.

(f) Hospitals utilized.

(d) Lawyer's office.

(g) Source of daily newspapers.

Fig. 32. Consumer preferences among Iowa farmers, 1934. (B.J.L. Berry and J. B. Parr, Market Centers and Retail Location: Theory and Applications. Englewood Cliffs, N.J.: Prentice-Hall, 1988, p. 12. Reprinted by permission of the publisher.)

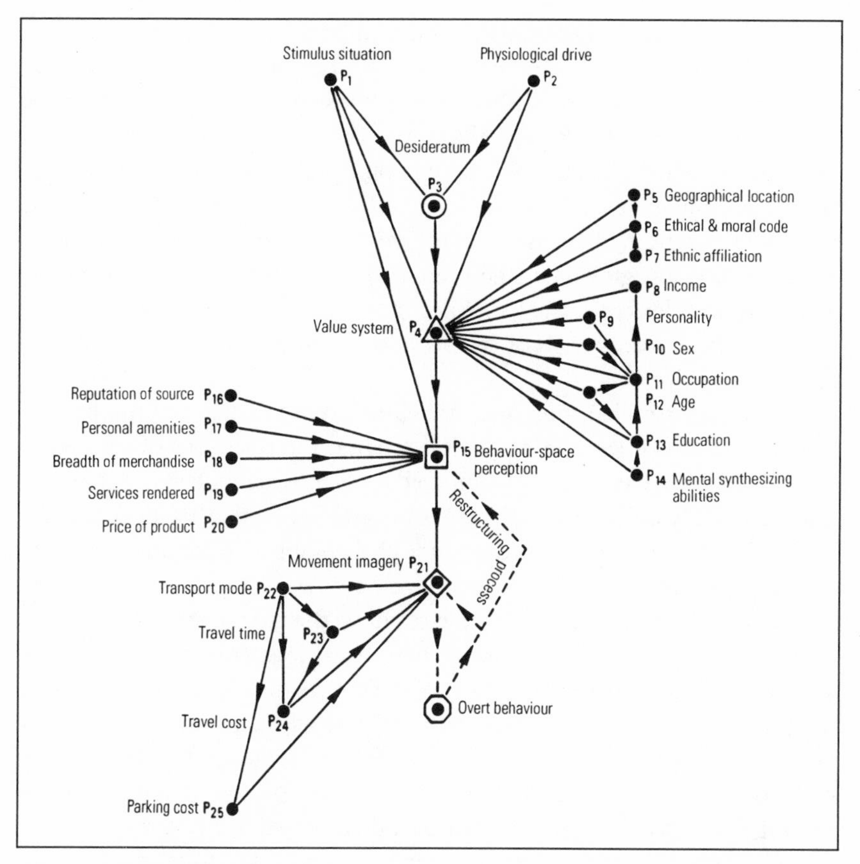

Fig. 33. Influences on consumer behavior. (Huff 1960, 165. Reproduced by permission of the Regional Science Association.)

and the construction industry. Critique of the potentially imperialistic consequences of applied mechanistic geography from the late sixties on came from a wide variety of sources. There were existentialist worries about the impact on daily life and thought, and other ideological doubts about the political power implications of a geography of domination; there was malaise, too, about the aesthetic and emotional impoverishment of landscapes shaped by standardized, interchangeable components.

A kaleidoscopic dance of metaphors was involved in the literature of critique—some contextualist, some formist, but a substantial number in the categories of mechanism itself, for example, from Marxist and structuralist sources. Systems were to be toppled by countersystems, theories by countertheories, the domination of one class replaced by another (Blaut 1970;

Harvey 1972). This in itself illustrates the enduring hold of the mechanistic root metaphor and its perennial ability to confront oppression with the language of the oppressor, rather than risk losing the cherished integrated perspective on reality. It also illustrates its weakness, for many of the grounds on which critique was raised could *not* be articulated in systems language (Samuels 1971; Buttimer 1974; Relph 1974). The 1980s again offered a Promethean challenge to mechanist metaphors as the international research effort on global change presented itself. With hitherto inconceivable technical devices for assembling, modeling, and analyzing biospheric and geological processes, *logos* and *ergon* reestablished their priority as professional practice.

In terms of milieu interests, mechanism certainly has had most to say on questions of order. It has paid less attention to questions of identity or niche (health or wholeness). Combined with mosaic, its favored *ergon* has unquestionably been to address issues of order. Now that issues of humanity's collective niche have again become challenging, and the analytical gaze has again begun to focus on interactions between biological and geophysical processes, one potential outcome could be the rediscovery of a mutually enriching dialogue with organism. There are at least two grounds for such speculation. First, there is the issue of scale horizons. It is on questions of horizon that mechanism, especially in its positivist expressions, faces difficulties. From the fifties on, mechanism in human geography has had the effect of concentrating research attention mainly on the national scale, reflecting the nature of available statistical data as well as the growing dominance of national sponsorship for geographic research. On selected themes, such as transportation and trade flows, mechanism reached toward regional and global horizons; however, for integrated studies incorporating more than one central aspect of life, mechanism appeared to function best within national boundaries. Second, the issue of how human activities are related to the physical environment, or how one might integrate physical and human aspects of life and landscapes, was rarely seen as either important or possible by economic and urban geographers in the period following World War II. In the heady enthusiasm of positivist economic geography, the physical environment was regarded ideally as a tabula rasa on which diverse models of behavior and interaction could be played. In the late eighties, systems-analytical language was again used to encourage joint research on global change by specialists on both sides of this tabula transformata (WCED 1987).

It is not only on the grounds of their complementary approaches to issues of order that mechanism and mosaic have again converged in the twentieth century. Their shared commitment to analytical rigor offers sound common denominators for debating questions of *logos*. In twentieth-century physics, the notion of *field* posed radical doubts about purely mechanical explanations of the universe (Einstein and Infeld 1938).

Field theory also revived organicist interest among psychologists of the Gestalt school, shedding new light on issues of communication within living forms (Bronowski 1960; Bertalanffy 1969; Haraway 1976; Phipps and Berdoulay 1985). Even if, for many physicists, mechanical explanations of the universe have far outrivaled those of form, others—for example, in biology—have sensed the need to reassert the reciprocity of these two perspectives. Sheldrake distinguished *formative causation* in morphogenetic fields from the "energetic type of causation with which physics already deals so thoroughly. Although morphogenetic fields can only bring about their effects in conjunction with energetic processes, they are not in themselves energetic" (1981, 71).[11]

Mechanism also shows signs of more openness toward contextualism and its operational theory of truth. It is beyond question that the biophysical and artifactual environment can be modeled and explained in mechanistic terms. As an elegant description of the status quo, and a formula for an autopsy of all that has become dysfunctional, few metaphors have proven more effective than mechanism. If one wishes to take the further step toward policy and normative action, then the systems-analytical mode of discourse tends to focus attention exclusively on managerial action, planning constraints, and proscriptions. The role of culture, place, and the physical environment in the formation of human identity and sustainable niche, however, can scarcely be subsumed under the traditional categories of mechanism. Beyond the delights of explanation and modeling lies the challenge of responsibility for future housekeeping and husbandry of the earth's resources: whether this is to be entrusted to global policing strategies or to jointly responsible communities at appropriate discretionary scales remains a contested ideological and political issue. Mechanism hitherto has revealed a preference for the first scenario, despite its overt commitments to emancipation and liberation. The discernment (or rediscovery) of workable solutions to resource management and sustainability will require a much closer look at the history of popular traditions, values, and practices—an arena where contextualism has proven more insightful than other approaches (see Chapter 6).

The perennial mechanists' dream of reaching causal explanations of terrestrial reality as a whole has, throughout the twentieth century, witnessed marvels of Faustian achievement. Ironically, as in the Promethean tale, that very technology assumed a life of its own, and virtually all conventional notions of cosmic order were challenged by the new science of chaos.[12] Yet it is largely due to advances in computer technology that the new discipline of chaos itself now promises deeper insight into mechanics. Even those mechanical systems that can be described very simply in deterministic terms can also demonstrate complicated and practically unpredictable behavior such as that associated with purely stochastic behavior. In this seemingly incomprehensible picture, certain distinctly recognizable structures emerge, as Prigogine and Stengers succinctly describe in their *Order out of Chaos* (1984). Truly random behavior seems to appear only within the realm of quantum mechanics and is therefore hardly observable except on a submicroscopic scale. Such a lack of determinism, however, plays an important part in astrophysics, making black holes—the formidable meeting point of general relativity and quantum mechanics—less black than previously anticipated (Hawking 1988, 99). The 1980s also welcomed the closely related notion of *fractals,* a concept that allows one to transpose scale upward and downward (Mandelbrot 1983). In reality, one could set limits in both directions, but the intermediate region shows a distinct fractal geometry. Using computer-based techniques, the geographer can now describe the shape of a tree, a lunar landscape, or the edges of Loch Lomond in mathematically precise terms. So much for Diderot's (1818) prophecy!

The quest for a complete theory of the universe continues to challenge the physicist. "Even if there is only one possible unified theory, just one set of rules and equations," Stephen Hawking asks, "what is it that breathes fire into the equations and makes a universe for them to describe?" (Hawking 1988, 174). As with Democritus, Lucretius, and the *Encyclopédistes,* the Promethean promise beckons: "If we do discover a complete theory, it should in time be understandable in broad principle by everyone, not just a few scientists. Then we shall all, philosophers, scientists, and just ordinary people, be able to take part in the discussion of the question of why it is that we and the universe exist. If we find the answer to that, it would be the ultimate triumph of human reason—for then we would know the mind of God" (175).

Chapter Five

World as Organic Whole

> Nature, considered rationally, that is to say submitted to the process of thought, is a unity in diversity of phenomena, a harmony blending together all created things, however dissimilar in form and attributes, one great whole animated by the breath of life.
>
> —von Humboldt 1848, 1:2

TO IMAGINE the world as a whole, seeking organic connections among its diverse components, has been the catalyst for creativity among artists, scientists, poets, and potentates throughout human history. Organicity in the cosmos and even in its terrestrial expressions has stirred geographical curiosity in many world civilizations. Traces of an organicist world-view can be found in the Vedic myth of Purusha (Zaehner 1966; McClagan 1977), in Chinese poetry and legend (Chung-yuan 1963), in pre-Socratic doctrines and the Hippocratic school of medicine (Kirk and Raven 1962; Glacken 1967), in Paul of Tarsus's letters about the Christian church as a mystical body, and in diverse cultural images of the world as cosmic tree.

Curiosity about relationships between humanity and nature has surfaced more among wanderers and explorers or visionaries of alternative styles of social and political life, peace, and justice than it has among academic researchers within their settled niches of expertise. For the ecologically minded today, those very bonds that cultures have forged with their environments constitute the most telling evidence of the Western world's need to rediscover a caring attitude toward nature:

> It is easy to underestimate the power of a long-term association with the land, not just with a specific spot but with the span of it in memory and imagination, how it fills, for example, one's dreams. For some people, what they are is not finished at the skin, but continues with the reach of the senses out into the land. If the land is summarily disfigured or reorganized, it causes them psychological pain. Again, such people are attached to the land as if by luminous fibers; and they live in a kind of time that is not of the moment but, in concert with memory, extensive, measured by a lifetime. To cut these fibers causes not only pain but a sense of dislocation. (Lopez 1986, 250)

The powerful appeal of organism as root metaphor of reality may be explainable in terms of its grounding in the most universal and intimate experience of all humans, that is, the experience of one's own body. Each body, composed of many distinct and unique parts, obviously functions as a whole. Each life journey, too, moves through stages and cycles, continuously dependent on reciprocal relations with other lives and milieus. Analogies to corporeal and social experience have allowed popular acceptance of such expressions as the "body politic," "arteries of circulation," "veins," and "limbs" (Barkan 1975). Myriad are the models and paradigms of research which spring from the image of society as organism (Phillips 1976, 49). Organic analogies have been used to describe the contours and nature of communities in place; they have also been used to explain the evolutionary processes within a society (Burrow 1966; Voget 1975). In the nineteenth century, few metaphors were "so striking or so compelling as the image of the social organism, of society as a living, self-perpetuating, integral and adaptable totality" (Coleman 1966). The classical texts of geography are replete with metaphors of circulation, maturation, metabolism, and synergy; the most powerfully unifying metaphor of all perhaps being that of the hydrological cycle (Tuan 1968, 1978).

Symbols based on the human body have inevitably been used in the elaboration of other world-views: its anatomy for formism, its physiology in mechanism, its uniqueness and individuality for contextualism. In organicism, however, the coherence and unity of reality is stressed; hence the favored (Renaissance) notion of the whole human person (body, soul, personality) as a microcosm of the universe (fig. 34).

> Man, like all living organisms, developed not only under the influences of terrestrial nature but of the cosmos. Physiological processes on earth are geared to cosmic forces; they reveal cycles that have daily, seasonal, and other periodicities that are linked to astronomical rhythms. Man is more than a child of the earth: he is a child of the universe to the extent that his body responds to cosmic rhythms and his mind finds enchantment and assurance in the stars. (Tuan 1971, 6)

Given its universal scope, it is not surprising that traces of organicism can be found in the records of virtually all world civilizations. Invariably, too, its expressions and applications have mirrored key cultural values and ongoing societal concerns. A deep-seated contrast has often been drawn, for example, between an "Occidental" penchant for dominating nature with explanatory models of thought, technology, and politics, and an "Oriental" one for developing sympathy with all living beings and a submissive role for humans within the orchestration of natural forces:

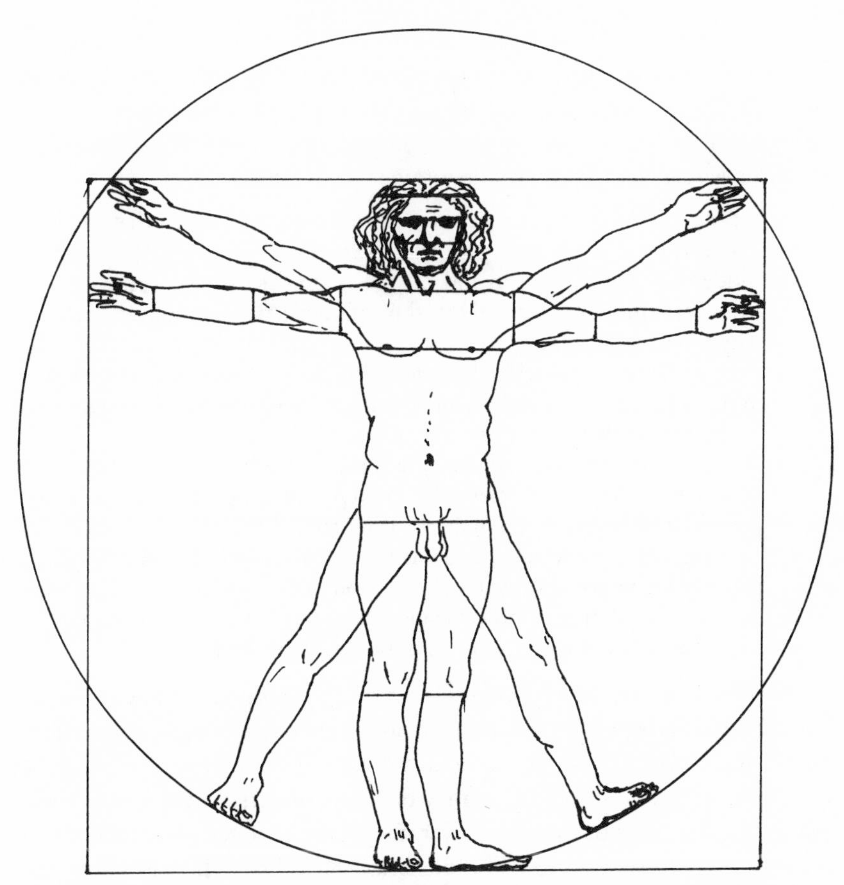

Fig. 34. Leonardo da Vinci's image of the human body as microcosm of the universe. (Sketch by Bertram Broberg.)

> I gather chrysanthemums at the eastern hedgerow
> And silently gaze at the southern mountains.
> The mountain air is beautiful in the sunset,
> And the birds flocking together return home.
> Among all these things is a real meaning,
> Yet when I try to express it, I become lost in "no-words."
> (T'ao Ch'ien, 4th c., cited in Chung-yuan 1963, 19)

Contrasts of Oriental and Occidental ideas, however appealing, become less credible when one explores the thought and life of individuals throughout history. The writings of Rhineland mystics during late medieval times, Bruno's *de l'Infinito Universo e Mondi* of the sixteenth century, Hölderlin's poetry of the nineteenth, all echo the refrain of human longing for closer sympathy and fusion with nature. Emerson's address to a Boston audience in 1833 concluded with the following words: "We feel there is an occult relation between the very worm, the crawling scorpions, and man" (Whicher 1953, 10). Across cultures and periods, organicist images of the world have displayed common denominators of concern for

1. wholes rather than parts—coherent unities so constituted that they involve more than an aggregation of the parts;
2. understanding dynamism and change among elements—structural transformations brought about through dialectical tensions, attractions, and repulsions;
3. transcendence—processes leading to higher levels of complexity and unity whereby previous contradictions and tensions could be resolved in higher-order integrations of reality; and
4. relationships between humanity and the orchestration of Being—a conviction that human elements could not be properly understood in terms of the dynamics of material processes or the "natural" determinisms of organic and inorganic life; for some, the quintessentially human elements of spirit (soul, emotion, will) provide the most satisfactory elucidation of how the universe as a whole was constituted, for others, the essential key lies in bioenergetic fields.

Of the four "world hypotheses" sketched by Pepper, organicism is perhaps the least amenable to description in exclusively cerebral terms. Animistic belief, mystical experience, and theological controversy permeate its historical expressions. Organism may be better understood as a metaphorical appeal for a global world-view rather than as a set of guidelines for systematic analysis or a rational explanation of reality. When attempts have been made to transpose its *poesis* into scientific theory (*logos*) or into rationalizations for political action (*ergon*), organicism has sometimes become an unwieldy monster (Buttimer 1985b). Its story in Western intellectual life could indeed be framed in terms of Phoenix, Faust, and Narcissus. Its acceptance or rejection at different moments in Western intellectual history may have had far more to do with emotional, aesthetic, and moral preferences than with epistemological or rational arguments.

The poetic appeal of "wholeness" played a Phoenix role at times and places when a liberation song about the integrity of life was called for—that is, whenever certain dimensions of experience, such as the social, intellectual, material, or spiritual aspects of human life, were ignored, oppressed, misconstrued, or simply forgotten. Giordano Bruno's sixteenth-century exploration into the potential infinity of the universe and mankind's rights and responsibilities for exploring new intellectual horizons was a liberation cry against a dogmatically circumscribed medieval world. Kropotkin's *Mutual Aid* (1902) protested against the post-Darwinian view that life was permeated by competitive forces of natural selection; instead, he emphasized those propensities of all living beings to cooperate with and complement one another. Humboldt, Reclus, Ratzel, Kjellén, and Edgar Kant all proclaimed the significance of emotional bonds that human groups had established with their home milieus—bonds that could scarcely be explained in terms of rationality, efficiency, or survival.

Once heard, such Phoenix songs would at times become absorbed in Faustian structures such as nations, unions, or empires; the poetic element submitted to a legitimizing *logos* or an expedient *ergon*. The ensuing structures would often eventually pose serious tensions between the original idea or *ethos* and the survival and maintenance of the institutions to which it had given birth. Organicism in its Faustian garb has bequeathed many embarrassing legacies: imperialism and colonialism, frontier conquest and destruction of the commons. Perhaps most poignant of all in the twentieth century are memories of the Holocaust and the expansionary *Geopolitik* of the Third Reich, which used organicist rhetoric for its own tragic ends. It has often taken generations or centuries before scholarly minds would dare to reconsider this world-view.

In historical perspective, too, one cannot fail to observe that organicist *poesis,* nostalgically treasuring the inspiration of its Phoenix days, inspired emotionally charged rhetoric against the excesses of Faustian eras—for example, in the philosophy and poetry of German romanticism and in the literature of New England Transcendentalists during the nineteenth century. The latter years of the twentieth century have again witnessed a potentially creative outcome from organicist strains in postmodernist literature, music, and art, which for virtually an entire generation has expressed Western humanity's malaise about scientific reductionism, environmental destruction, and the tyranny of mechanistic definitions of reality.

Classical Foundations of Organicism

The origins of organicist thought in the West can be traced to the pre-Socratics. The Mediterranean world itself, meeting ground of diverse civilizations, in all its geodiversity, no doubt stirred inquiry about natural and human phenomena. Speculations about nature and order in the universe were also combined with more practical concerns about health and disease, trade and travel, politics and morality, and the significance of place, space, and time in human life. The Mediterranean world remains the best illustration of intimate connections between geography and human interests, identity, order, niche, and horizon (Semple 1931; Sion 1934; Braudel [1966] 1972).

Heraclitus (ca. 540–ca. 480 B.C.) taught that the essential human challenge was to understand the *logos,* or universal formula underlying all natural processes. Logos manifested itself for him in the connection between opposites: changes in one direction were ultimately balanced by corresponding movements in the other. Order in the world was "an ever living fire kindling in measures and being extinguished in measures," as for instance, relationships between sea and land. The persistence of unity through apparent change was succinctly affirmed in his famous slogan, *Upon those who step into the same river different and ever different waters flow down*. While skeptical about conventional religious rites, Heraclitus identified a God with those opposites that persisted throughout their changes and was thus capable of understanding them perfectly (Kirk 1974; Kirk and Raven 1962).

Empedocles (492–432 B.C.) identified the four elemental "roots" of reality (fire, air, earth, and water) upon which the two opposing forces of love (unifying) and hate (divisive) reacted (Sambursky [1956] 1987). Few theories have enjoyed as much acclaim as this doctrine of the elements and their counterparts in the "humors" of the human body. Just like analogues based on the human body itself, the "elements" have been used in both formist and mechanist world-views (see Chapters 3, 4). But organicists, for example, in the Alcmaeon school of medicine (ca. 500 B.C.) sought here some clues to the mystery of health and disease in the proper proportioning and balancing of these humors.[1] Hippocrates taught that "the human body contains blood, phlegm, yellow bile and black bile; these are the things that make up its constitution and cause its pain and health. Health is primarily that state in which those constituent substances are in correct proportion to each other, both in strength and quantity" (*Nature of Man,* 4: 262).

Since good health demanded a proper proportioning of the humors, and since certain aspects of the physical environment such as temperature or moisture could tip the balance among them, it was felt that the dominance of one or the other would vary among climatic zones, even among different seasons of the same zone.

Greek curiosity was certainly not confined to terrestrial phenomena. Observations of movement among heavenly bodies led to speculations about order in the universe as a whole. Such astronomical curiosities could almost be expected in Mediterranean islands and shoreline cities (Semple 1931); seasonal changes could be associated with the appearance of certain bodies in the skies. Was it not plausible to suspect that the stars exercised an influence on human life too? Some scholars trace the origin of astral religion to the Chaldeans in the sixth century B.C. (Cumont 1959), and ascribe analogues between the structure of the cosmos and those of human society to Sumerian thought (Kramer 1959). In late Greek and Roman times, astrology became the source of organicist conceptions of the universe, "a cosmic environmentalism" (Thorndike 1955). Glacken claims that "the idea that there is a unity and a harmony in nature is probably the most important idea, in its effect on geographical thought, that we have received from the Greeks, even if among them there was no unanimity regarding the nature of this unity and harmony" (1967, 17).

With the refinement of analytical approaches to the study of nature in post-Socratic times, one could also note a change of metaphorical vision. From the dialectical and dynamic conception of opposing forces, which was central to pre-Socratic writers, one already finds in Plato and Aristotle a tendency to seek balance, harmony, and formal correspondences among phenomena. There is also a metaphorical transformation from organicist to formist (and eventually to mechanist) ways of construing nature and health (Hudson-Rodd 1991). The poetic and heuristic tones of Anaximander, Heraclitus, and Empedocles became transposed to logically analyzable and practically manipulable categories by the Socratics and later by the Stoics. "Conflict between opposing forces" became transposed to "harmonic blending of the powers," and formal correspondences between earth, heaven, and human society became a more engaging concern than the potential interplay of energies among them. The subsequent record of environmental determinism in the West might also be elucidated in these terms: through eliminating such intangible elements as "love" and "hate" or any other transcendent force in the relationships between humanity and its milieus, whether in formistic or mechanistic accounts of human society and environment, deterministic interpretations were rendered plausible. It would take counterarguments of an organicist or contextualist kind eventually to confound them.

Although these doctrines of Greek origin may have impressed themselves most effectively on the Western imagination, it is to Asian sources that one should turn for the richest sources of inspiration for an organicist world-view. Few of its features have caused greater controversy and confusion than its insistence upon unity in diversity—the One in the many, the Whole in the parts (Sällström 1986; Duncker 1985). A radical difference between European and Asian varieties of organicism might be captured in the distinction between what one could label "cosmic" and "mayan" foundational myths. The Greek word *cosmos* meant order to be imposed on *chaos;* it implied a unified order as experienced in social, military, and ceremonial life. It was probably in Pythagorean times (ca. 530 B.C.) that the term was applied to nature and the physical earth. The Sanskrit word *maya,* on the other hand, denotes an endless proliferation of manifest forms in the universe, and it connotes measure and power. Both cosmic and mayan myths imply an evolutionary sequence of creation, and a hierarchy of forms in which humankind is almost universally considered as climax. The contrast is most probably ascribable to radically different teleological interpretations of the creative process. Was there one transcendent *omega* point toward which all evolution directed itself, as monotheistic religions believed, or were there many forms, each proceeding according to its own intrinsic nature, fulfilling itself within a pluralistic universe, as pantheists might believe? Organicism in the West has expressed preferences for both cosmic and mayan approaches to reality but has characteristically chosen the former.

In terms of human knowledge and understanding, organicism has maintained a closer kinship with the so-called paradoxical logic of the Orient than with the "categorical" logic of the post-Socratic West (fig. 31). "Words that are strictly true seem to be paradoxical," Lao-tse ("the old Master") taught, and Heraclitus complained, "They do not understand that the all One, conflicting in itself, is identical with itself: conflicting harmony as in the bow and in the lyre." Paradoxical logic ultimately leads toward an understanding of how much one does not know; categorical logic leads toward a sense of knowing reality conceptually. Brahmanic philosophy concerned itself with the distinction between the manifoldness of phenomena and their ontological unity. Harmony (unity) consisted in the conflicting positions from which it was constituted; hence, reflections were centered on the paradox of simultaneous antagonisms of the manifest forces and forms of the phenomenal world. The ultimate power in the universe was regarded as transcending both conceptual and sensual spheres (Aristotle's *kosmos noétos* and *kosmos aesthétos*). In Brahmanic philosophy, perceived opposites reflected not the nature of things but the nature of the perceiving mind (fig. 35).[2]

The alleged contrasts of Orient and Occident are less apparent when one examines geographic thought and practice. In the *I Ching* one finds analogous attempts to identify elements and to explain natural processes, for example, the hydrological cycle (fig. 36).

With some notable exceptions among mystics and rebels of the Western tradition, the first millennium of Christendom allowed little room for organicism and its paradoxical logic. Medieval times, often regarded as the Age of Belief (Fremantle 1954), nurtured a basically formistic conception of the universe (Chapter 3). When Platonic and Aristotelian intellectual traditions eventually converged in the *studia generalia* of northern Italy and the Iberian Peninsula, an Age of Adventure was born. Organicist strains again

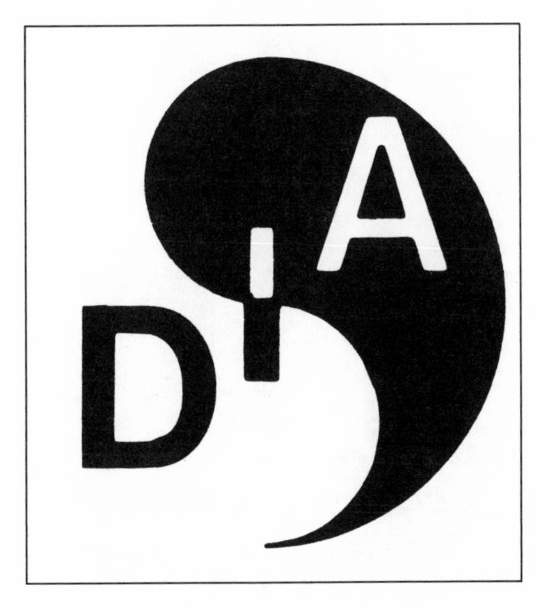

Fig. 35. Yin/Yang: logo of the International Dialogue Project

resounded during that *aggiornamento* which the Mediterranean world experienced in the late medieval and early Renaissance period.

An Infinity of Worlds

The millennium stretching from the fourth to the fourteenth century has often been depicted as the Dark Ages, during which ecclesiastical authority and feudal modes of social control exercised their sway over servile human populations. It was, however, also an age when organically cohesive *genres de vie* were developed within geographically circumscribed communities (Dubos 1972; Bookchin 1974). Certainly it was an age innocent of the power politics of nation-states and the babel of tongues which later generations experienced as a primary impediment to organicity in the world of thought. Latin provided a universal vernacular for scientists, at least, even if in the long run this played a conservative role in geographic thought and practice (de Dainville 1941; Broc 1986). There are elegant accounts of organicity in the daily liturgies of Benedictine monasteries, of attunement to clock time and daily rhythm which may eventually have provided the basic discipline required for the Industrial Revolution (Mumford 1934). The organicity achieved in medieval communities was the kind that was possible within a "closed system": one that later generations might find too confin-

Trigram Symbol Objects Represented

- *Heaven*, king, father
- *Earth,* official, mother, wife
- *Thunder,* eldest son
- *Wind, wood,* eldest daughter
- *Water,* channel, pit, middle son
- Fire, sun, lightning, middle daughter
- *Mountain,* path, little stone, youngest son
- *Collection of water, pond,* youngest daughter

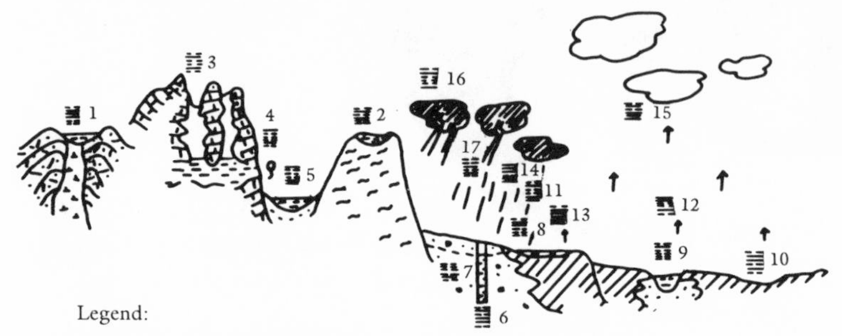

Legend:

1. "there is a pond on the mountain"
2. "there is water on top of the mountain"
3. "water continually comes out"
4. "springs originate under the mountains"
5. "there are lakes under the mountains"
6. "water is present on the wood"
7. "water exists in the ground"
8. "there is a pond or lake on the ground surface"
9. "water exists on the ground surface"
10. "there is no water in the pond" (viz., the pond is dried up)
11. "water exists over the pond" (viz., it rains over the pond)
12. "wind blows over the water" (evaporation)
13. "the pond ascends into the sky" (This means that almost all the water of the pond has evaporated)
14. "the sky and water leave each other" (It is raining)
15. "clouds ascend up to the sky"
16. "cloud and thunder"
17. "thunder and shower"

Fig. 36. The *I Ching* and the hydrological cycle. The *Book of Changes,* one of the oldest Confucian classics, was used for divination. It is based on eight trigrams, combining to create 64 hexagrams, each comprising two trigrams. Each trigram is formed by three solid or broken lines, and all are associated with a constellation of meaning. Object representations used in fig. 36 are underlined. Although the term *water cycle* is not used in the *Book of Changes,* a combination of 17 hexagrams, in which one or both trigrams refer to water or a collection of water, affords a comprehensive account of the hydrological cycle. (After Shen 1985, redrawn by Xiaopeng Shen.)

ing and static, as but one level of integration in humanity's way of relating to its terrestrial home.

The Phoenix esprit of Renaissance times was ushered in by visionaries with world-views of distinctly organicist horizons, closer in spirit to the pre-Socratics than to their medieval mentors. Though initially obliged to articulate their messages in formist categories, and indeed delivering marvels of form in their artistic and architectural works, their fundamental images of reality could not be confined to those of formism. Inevitably, their innovations generated violent reactions from the guardians of orthodoxy or from the potential builders of new orthodoxies. Two highly contrasting and mutually opposed world-views confronted each other: organicism and formism. The world-view of mechanism offered a plausible alternative: like organicism, it presented an integrated picture of reality; like formism, it rested its cognitive claims on analytical rather than synthetic approaches to knowledge (see Chapter 3).

For Renaissance pioneers, nature's work of art could be understood from analogies between one's own body and that of nature as a whole (Barkan 1975; Mills 1982, 242). The speculations of Bruno, Paracelsus, and Leonardo da Vinci opened spaces beyond the terrestrial sphere, as Koyré has so eloquently described in *From the Closed World to the Infinite Universe* (1957). Nicolas of Cusa may have pioneered relativistic conceptions of space and time (see Chapter 6), but his disciple, Giordano Bruno, was the poet of the infinite universe (Rocchi 1989). "In every man," Bruno taught, "a world, a universe, regards itself." And its creator could indeed have initiated the process whereby matter-in-motion could co-create an endless variety of worlds: "Thus is the excellence of God magnified and the greatness of his kingdom made manifest; he is glorified not in one, but in countless suns; not in a single earth, but in a thousand, I say, in an infinity of worlds" (cited in Koyré 1957, 42).

Fifteenth-century European scholars, availing themselves of scientific insights developed in the Arab world, reintroduced dynamism and change in the world picture (Glacken 1967, 254–84; Mills 1982). Rather than being a pyramid of hierarchically arranged essence subject to a "natural" law devised by some extraneous, albeit benign, divine artisan, the cosmos would now be regarded as possessing life, intelligence, and soul; it would be seen as passing through stages of infancy, youth, maturity, and old age; it had skin, hair, a heart, a stomach, veins, and arteries. From the vantage point of physical geography, Mills noted, there were two enabling aspects of this Renaissance image. First, the extraction of precious stones was considered somewhat like abortion; for it was believed that all minerals, left to themselves, would eventually "ripen" into gold (1982, 244). Burning curiosity was also evoked about the hydrologic cycle and the role of water in the workings of the earth as a whole. Leonardo wrote: "The body of the earth, like the bodies of animals, is intersected with ramifications of veins which are all in connection and are constituted to give nutriment and life to the earth and its creatures. These come from the depths of the sea and, after many revolutions, have to return by the rivers created by the bursting of these veins high up" (1970, 2: 62; cited in Mills 1982, 244). But how was water conducted from the oceans to the tops of mountains? Leonardo speculated:

> The waters circulate with constant motion from the utmost depths of the sea to the highest summits of the mountains, not obeying the nature of heavy matter; and in this case they act as does the blood of animals which is always moving from the sea of the heart and flows to the top of their heads; and he who bursts veins—as one may see when a vein bursts in the nose, that all the blood from below rises to the level of the burst vein. When the water rushes out of a burst vein in the earth it obeys the nature of other things heavier than air, whence it always seeks the lowest places. (1970, 158).

These poetic speculations eventually resonated in the eighteenth-century discovery of the blood circulation system by Harvey and in Hutton's *Theory of the Earth* (Hutton 1795; Nicolson 1960; see also Chapter 4). Leonardo da Vinci remains the model of "Renaissance Man": he mapped and measured, experimented and invented, all the while passionately curious about the workings of nature (Gibbs-Smith 1985). As in the case of Galileo, however, organicist elements were replaced by mechanist ones, and it was as pioneers of mechanism that both men were applauded in Western history. The Faustian harvest was eventually reaped during the ages of Reason and Enlightenment, which stretched from the late sixteenth through the eighteenth centuries, times when mechanism and formism held sway in the intellectual drama. By the seventeenth century, the most pervasive tension in scholarly thought about humanity and nature was that between organic and mechanical views. Glacken summarized:

> In the mechanical, the actions of the individual parts of a whole are explained by known laws, the whole being the sum of the parts and their interaction. In the organic, the whole exists first, perhaps in the mind of an artisan, before the parts, the design of the whole explains the actions and reactions of the parts. The mechanical view, emphasizing secondary causes, eliminated final causes as active guides in investigation, relegating them to theology or private piety. In the seventeenth century writings on natural law, matter and motion, one frequently reads attacks on Democritus, Epicurus, Lucretius, by partisans of final causes who are, moreover, interested in appropriating to their own uses the new knowl-

> edge embodied in natural law. Classical authors are often cited as authorities against the "mechanical ideas," and Plato, Seneca, and Cicero are constantly making depositions. Was nature a system based on law, or providentially designed, or the product of design for a purpose? (Glacken 1967, 378)

The contrasts between mechanical and organic world-views have perhaps been overplayed. Schlanger claimed that these differences were never so sharply drawn in France as they were in Germany (Schlanger 1971, 59). On issues of lawfulness, of design and purpose in the universe, monotheists, pantheists, and agnostics pursued different courses; the debate within physico-theology was conducted within the frameworks of formism and mechanism (see Chapters 3, 4). Organicists rose above these battles of orthodoxy, affirming the plausibility of a transcendental force in the universe as a whole, their primary concern being the coherence and unity of reality, not its particular constituents. This mystique of wholes (*Gestalten*), coupled with a sensitivity to spiritual as well as material energies in nature and in human history, characterized romanticism, the era of Kant, Hegel, and Goethe, Coleridge and Wordsworth. This is also the period when the foundations of geography as a discipline were established in German universities. And already in the pioneering works of Alexander von Humboldt (1769–1859) and Karl Ritter (1779–1859), the tension between two fundamentally opposed world-views, organism and mosaic, was often evident. Ritter's *Erdkunde* led the way for encyclopedic regional accounts fashioned within the metaphorical mode of transcendent formism. Humboldt's *Cosmos,* however, remains the prototype of organicism.

Cosmos, Human History, and Geography

Just as Renaissance scholars wished to emancipate themselves from the strictures of a formist world-view, so also scholars of the romantic period sought liberation from the mechanical certainties of Enlightenment rationality. Two eighteenth- and early nineteenth-century writers stand out as particularly influential in the subsequent progress of organicism in geography: Johann Gottfried von Herder (1744–1803) and Alexander von Humboldt. Both were highly influenced by Goethe and, like him, were skeptical about the rationalistic and mechanical interpretations of nature and society proclaimed by their French colleagues of Cartesian inspiration. Goethe criticized the traditional Western penchant for ascribing teleological purposes to nature, especially the notion that everything was created for human purposes. "In Goethe's hands," Glacken observed, "the environment is metamorphosed from a passive medium of life to an active creative role in conditioning and maintaining it."

Herder, like Spinoza (1632–77), claimed that for the universe to be intelligible it should be conceived as a whole, as an eternal, self-creating, self-explanatory system (Hampshire 1956, 105). Far from being a great machine, explainable in terms of mechanical laws, the world was regarded by Herder as a "living organism" made up of "dynamic, purpose-seeking forces, the interplay of which constitutes all movement and growth" (Berlin 1976). Like Leibniz (1646–1716), he regarded the universe as constituted of ultimately irreducible substances (monads) in the form of individual persons, whose individuality and integrity emerged from their participation in larger wholes: "Every man is ultimately a world, in external appearance indeed similar to others, but internally an individual being, with whom no other coincides" (Herder 1784–91, bk. 7, chap. 1).

Likewise each social group, or *Volk,* developed its individuality and integrity through the dialogue of human ingenuity and local environments. The history of the world was the story of diverse *Völker* and their record of ingenuity and imagination as expressed in myth, metaphor, religion, and ways of life. Linear accounts that conventionally placed European civilization at the pinnacle of progress failed to do justice to time, place, and national character in the lived reality of people's daily existence. Herder was especially agnostic about faith in scientific progress then espoused by French *Encyclopédistes.* Of conventional typologies of societies and states, such as Montesquieu's trilogy of monarchial, aristocratic, and democratic, he was highly critical; he wished to confound the environmental determinism of Buffon and Montesquieu, acknowledging, however, the latter's insights into connections between climate and social life (Montesquieu 1950). Mankind is intimately dependent on natural resources, Herder admitted, but it would be ludicrous to "compare him to the absorbing sponge, the sparkling tinder; he is a multitudinous harmony, a living self, on whom the harmony of all the powers that surround him operates" (Herder 1784–91, bk. 7, chap. 1).

History implied continuous movement and change, but catalytic moments for each Volk were the discovery of religious or linguistic identity. Herder claimed that there was an optimal time sequence in the development of Volk potentialities (Herder 1784–91, bk. 4). From the infinite variety of possible growth times, the most appropriate rate should be determined by the nature of the environment and the "original genius" of a people. Cultures passed through cycles of infancy, maturation, and death; some even "outlived" themselves, as in the case of Rome, "dragging out their monotonous existences, stinking corpses."[3] Human history should not therefore be based on the evidence contained in state and government records, but rather on the music, literature, art, and science (in that order) of diverse peoples. The historian should seek to transpose himself into the lived situations of the actors, to understand and communicate the coherence of a particular way of life, feeling, and action (Berlin 1976).[4]

Echoes of this ecumenical attitude toward world civilizations certainly inspired the explorations of the Forsters and that curiosity about "primitive" peoples which characterized much of the literature of romanticism (Forster 1777). The Forsters' accounts were harbingers of an era of scientific travel and exploration, and they certainly inspired Alexander von Humboldt (Glacken 1967, 501–3). Underlying Herder's cultural relativism, however, there was a teleological tone, no doubt reflective of Protestant pietism, which suggested that history was predetermined by a supernatural deity: "Ages roll on, and with them the offspring of ages, multiform man. Every thing that could blossom upon Earth has blossomed; each in its due season, and its proper sphere: it has withered away, and will blossom again, when its time arrives. The work of Providence pursues its eternal course, according to grand universal laws" (Herder 1784–91, bk. 14, chap. 6).

These grand universal laws, which Herder sought to elaborate in the context of world history, were explored by Karl Ritter in the regional geography of the world. The earth was designed as the playground or nursery (*das Erziehungshaus*) for humankind; its physiognomy bore the marks of a providential plan, which geography as a discipline could explore: "Erdkunde should strive to embrace the most complete and the most cosmical view of the earth, to sum up and organize into a beautiful unit, all that we know of the globe. . . . It should also show the connection of this unified whole with man and his creator" (cited in Taylor 1951, 44).

Preamble aside, the diverse works of Karl Ritter scarcely qualify as expressions of an organicist world-view. Methodologically and philosophically, Ritter offers a better illustration of the Kantian prospectus for geography, namely, formal description of the earth's surface. In Humboldt's work, he most admired the zonal mapping of vegetation (1806). On priorities for disciplinary practice, however, he placed regional synthesis far above systematic inquiry (see Chapter 3).

Alexander von Humboldt, unlike Herder and Ritter, saw no need to postulate an extraterrestrial guiding force in nature and human history. "The principal impulse by which I was directed was the earnest endeavour to comprehend the phenomena of physical objects in their general connection, and to present nature as one great whole, moved and animated by internal forces" (Humboldt, *Cosmos,* 1: vii). Later in the same work, he reiterated, "Nature is not dead matter. She is, as Schelling expressed it, the sacred and primary force" (379). With a breadth of vision and ingenuity which could span the two great intellectual movements of his century, the Enlightenment and romanticism, Humboldt integrated scientific and humanist approaches to nature: "May my descriptions render to the reader a part of the enjoyment which a susceptible mind finds in the direct observation of nature. Because enjoyment grows corresponding to an increasing recognition of the inner coherence of the natural forces, I added to each article scientific explanations and additions" (Humboldt, *Cosmos,* 5–8: 1–2, 13, 283).

Impressed, as Goethe was, with the analytical potential of Enlightenment science, and equally convinced of the values proclaimed by its romantic critics, Humboldt fashioned a view of the world which transcended the limits of mechanism and formism. The study of nature demanded more than exhaustive classifications of genera and species, à la Linnaeus; still more, it needed to be understood as infinitely more complex than a system of mechanical laws:

> The catalogues of organized beings to which was formerly given the pompous title of "Systems of Nature," present us with an admirably connected arrangement by analogies of structure. . . . But these pretended systems of nature, however ingenious their mode of classification may be, do not show us organic beings, as they are distributed in groups throughout our planet according to their different relations of latitude and elevation above the level of the sea and to climatic influences which are owing to general and often very remote causes . . .
>
> The distinction which must necessarily be made between descriptive botany (morphology of vegetables) and the geography of plants, is that in the physical history of the globe, the innumerable multitude of organized bodies which embellish creation are considered rather according to "zones of habitation" or "stations" and to differently inflected "isothermal bands," than with reference to the principle of gradation in the development of the internal organism. (Humboldt, *Cosmos,* 1: 42–43)

Organicity for Humboldt was a feature of the cosmos, not just simply of Planet Earth and its diverse regions or inhabitants. None of the fascinating particulars—of the High Andes, the continental plains, or island peripheries—could make sense until they were all placed within this cosmic horizon. In many ways, Humboldt articulated the cosmopolitan vision of early nineteenth-century German idealism, echoing the emancipatory fervor of Hegelian philosophy and the freedom of spirit celebrated in the works of Fichte and Schelling. In his commitment to science, however, he tempered the excesses of idealism: art and science were united in his approach (Bunksé 1981). His work received acclaim in France and America from painters and writers, scientists and historians. Ralph Waldo Emerson regarded him as "one of the wonders of the world, like Aristotle, or Julius Caesar . . . who appear from time to time, as if to show us the possibilities of the human mind" (1869). Frederick E. Church's landscape paintings, inspired by Humboldt, were displayed in the museums of London and New York in the 1850s (Bunksé 1981, 1990).

During the nineteenth century, however, the cosmic horizons and synthetic power of organicism were undermined. Even among Humboldt's own compatriots of the early 1800s, the Napoleonic conquest seemed to turn many a previous cosmopolitan into an ardent nationalist (Leclercq 1963, 25). Müller (1779–1829) wrote that "a nation forms a living being," and Johann Bluntshli (1808–81) considered the nation-state as "an organic whole, a moral-psychological organism, capable of synthesizing the ideas and sentiments of a people" (Leclercq 1963, 27). From 1823 on, the *Zeitschrift für Völkerpsychologie* published essays on the psychology of different peoples. Post-Darwinian science looked for more analytically incisive meanings of the word *organism,* thereby releasing heated controversies over environmental determinism and natural law. The metaphor itself assumed three distinct connotations: societal, biological, and ecological, all three recurrent in twentieth-century debates among sociologists and anthropologists. It was on early nineteenth-century examples, and particularly on Hegelian notions of the organic whole, that Pepper based his definition of this root metaphor and its claims to truth.

A Coherence Theory of Truth

An organicist world-view, in Pepper's scheme, sees every event in the world as more or less concealed process. It envisions world reality as, first, a scattered array of *fragments* which are attracted or repelled by one another until they congeal around *nexuses*. Through continuous dialectical processes of conflict and competition, these nexuses eventually become integrated into a higher unity called the *organic whole*. A distinction is drawn between so-called materialist and idealist interpretations of this organic whole, each selecting its own range of analytical categories. Pepper then differentiates between progressive and idealist categories:

Progressive categories of organicism are:

1. *fragments* of experience, which appear in
2. *nexuses,* or connections, which spontaneously lead, as a result of the aggravation of
3. *contradictions,* gaps, oppositions, or counteractions, to resolution in
4. *an organic whole.*

Ideal categories postulate that this organic whole is found to have been already

5. *implicit* in the fragments, and to
6. *transcend* the previous contradictions by means of a coherent totality, which
7. *economizes,* saves, preserves all the original fragments of experience without any loss. (S. C. Pepper 1942, 281–308)

According to Pepper, materialists assented to the first four (progressive) categories, but rejected the last three. Idealists concentrated on the latter three, sometimes neglecting the material categories. Both, however, are needed if the metaphor as a whole is to be cognitively credible. Without the idealist teleology associated with a Hegelian account of history marching from maximum fragmentation to ultimate integration, this world hypothesis loses its coherence. Organicism rests its cognitive claims on a *coherence theory of truth*. The Hegelian dialectic of thesis-antithesis-synthesis occurs at each stage in the drama of thought as well as life. Each fragment is forever restless in its "abstractness" until driven by its nexus to its exact opposite, which is also its complement. This dialectical process resolves itself into an integration, a higher-level synthesis, which recognizes the claims of all fragments, transcends them, and harmonizes them in a richer and more concrete whole. The process continues until there are no more nexuses flying around in search of satisfactions, but all are tied together within one absolutely coherent organic whole, namely, the Hegelian Absolute (fig. 37).

Pepper's account, focusing as it does on cognitive claims and illustrating these with examples from Hegelian times, leaves the metaphor *organism* largely shorn of its heuristic appeal. Its association with Bismarckian politics, with imperial expansionism, racism, and environmental determinism, has rendered it repulsive to the democratically inclined. Dogmatic adherence to creed, both materialist and idealist, has also served to repel the empirically oriented scholar. The appeal of an organicist world-view, its acceptance or rejection, cannot be satisfactorily explained in terms of its alleged epistemological or cognitive claims. One finds elements of organicism in the work of anarchists and conservatives, positivists and antipositivists, scientists and humanists of the nineteenth century. Much of its appeal can be ascribed to the material circumstances in which it was articulated and to the emancipatory role that its rhetoric played in nineteenth-century Euro-American contexts.

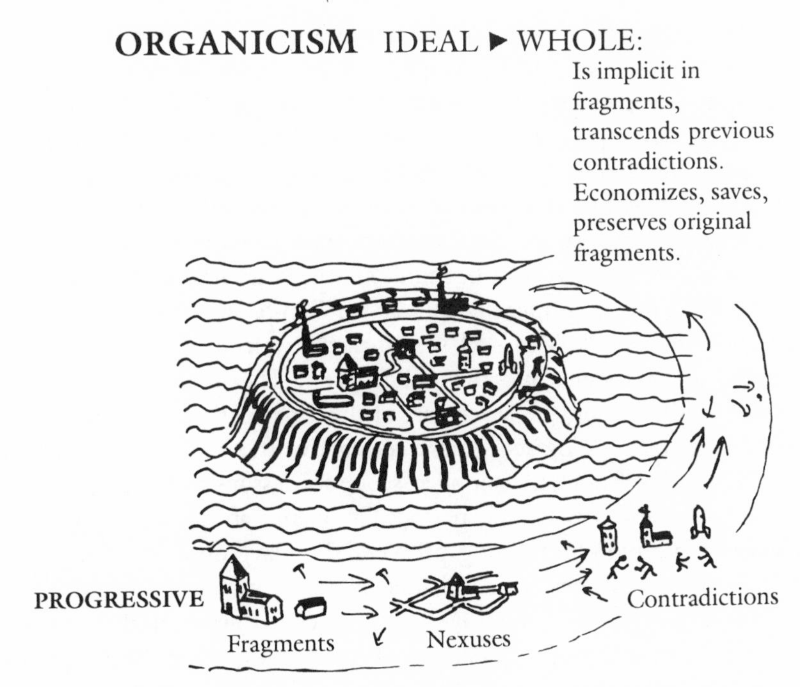

Fig. 37. A coherence theory of truth. (After Pepper 1942; sketch by Bertram Broberg.)

The Appeal of the Organism

Nature and Heimat

The eighteenth century was filled with revolutions, scientific and political; many scholars and artists sought refuge from mechanical certainties and found fresh sources of wisdom in nature. France had pioneered in designing rational means of implementing democracy (liberty, equality, and fraternity), and Germany had proclaimed universal (spiritual) humanity. Within a generation, however, the Napoleonic Wars had engendered a vigorous burgeoning of nation-states and national cultures, which often rested their identity claims on either the distinctiveness of their *Heimaten* (home territories) or the expansionary dreams of empire. Rhetorically speaking, organicism played a Phoenix role for both: for the former, as a symbolic guide for reemergence from situations of perceived oppression; for the latter, as legitimization for further conquest or noblesse oblige. Across national lines, revolutionary movements expanded and contracted, often begetting constellations of submovements and sects in political, religious, and civic life.

In the wake of Crimean and American wars, Europe witnessed a profound transformation of its landscapes with the growth of industry and the expansion of big cities. A Faustian myth of progress not only hastened the rational development of technology and economic growth within the metropole but also moved millions of adventurers, visionaries, geniuses, and landless peasants to explore the pioneer fringe across the Atlantic. A vast curiosity about nature itself was associated with this in home regions, not only as an object of scientific inquiry in the years after Darwin, but also as a challenge for geographical exploration. Tien-Shan or Gobi, Northwest Passage or the South Seas—all were heartily welcomed by geographical societies, governments, and the public press. The natural sciences progressed rapidly during the latter part of the nineteenth century, while historians and anthropologists debated alternative theories of human evolution. Nature itself not only supplied scope for lessons to be taught to the schoolchildren of crystallizing nations; it offered an attractive recreation ground for a rapidly urbanizing people and a playground for imaginative painters, poets, and musicians still attached to the values of romanticism.

Not that "nature" was always as munificent as urban aesthetes would sketch it. In most European societies of the day, a predominantly agrarian populace faced unprecedented challenges to its historical traditions, to its ways of thought and its ways of life. In many cases there was poverty, emigration, structural reform, a reshuffling of prestige and authority, and an invasion of external influences (e.g., commodity markets) into formerly self-sufficient or reasonably autonomous modes of dwelling. And the growing numbers of industrial workers, landless and unemployed, were becoming more articulate in their demands for justice and equality.

Common to most national settings, too, was a drive toward universal education. School curricula were improved and a fixed school-leaving age imposed. In Nordic countries, adult education programs, farmers' cooperatives, and folk high schools appeared, and a general will for self-improvement throughout all sectors of society became manifest in law and practice. Geography became an essential element in this move toward universal education. It was to address itself to the needs of the day, claiming an indispensable role in the edification of young citizens.

Each European country faced a somewhat different constellation of challenges at the end of the nineteenth century as geography made its claim for status in school and university curricula. At the risk of overgeneralization, one could say that France was busily implementing reforms in the wake of its revolution, Germany building the spirit of nationhood, England establishing its claims to an empire on which the sun would never set, and America creating a frontier society. The notion of organism demonstrated a remarkable plasticity in its adaptation to all these national interests. One seeks in vain to associate this metaphor exclusively with creed or culture, personality or period. The appeal of organicism and its expressions during the early years of the discipline of geography could not be fully understood without reference to the contexts in which it was articulated. A closer look at the material and cultural circumstances of the late nineteenth and early twentieth centuries can shed light on the similarities and contrasts in disciplinary practice on either side of the Atlantic.

Organicism and the Foundations for Geography

At the dawn of geography's establishment as a discipline in European universities during the latter half of the nineteenth century, organism had already begun to assume literal as well as metaphorical meanings, not only among scientists but also in popular parlance. While the founding fathers of the discipline found organicist rhetoric to be effective in convincing sponsors and audience, its applications in practice were fraught with logical and epistemological difficulties (Stoddart 1967). Traces of organicism, in the more global and poetic sense, were already discernible in the writings of historians during the nineteenth and early twentieth centuries. Herder's environmentally sensitive and cyclical interpretations of folk rather than elite history are echoed in Michelet's *Histoire de France* (1833–67), in Elisée Reclus's *La terre et les hommes* (1877; see also Dunbar 1981), and in Kropotkin's various works (1885, 1898, 1902; see also Breitbart 1981). Reclus wrote: "Each period in the life of mankind corresponds to a change in its environment. It is the inequality of planetary traits that created the diversity of human history. Life corresponds to the environment. The earth, climate, habit of work, type of food, race, kinship and the mode of social grouping: these are the fundamental facts which play their part in influencing the history of every man" (1877, 42).

Organicist approaches to social geography, such as those implicit in Herder's *Idéen,* inspired many radical scholars. Liberals, too, such as those of the Vidalian School of *la géographie humaine,* fashioned sensitive accounts of *genres de vie* in diverse milieus with a philosophy of history remarkably akin to that of Herder (Buttimer 1971).[5] "La France est une personne" (France is a person), Michelet declared (1833, 29); in fact, it had a dual personality, those of the peasant and the industrial worker. This was what distinguished it from Germany and England, countries that had been profoundly influenced by emigration (Michelet 1833). Later, Vidal de la

Blache adopted the term *personality* in his descriptions of France and its regions (Vidal de la Blache 1903). He was highly critical, however, of such notions as the nation-state as living organism, as expressed in Friedrich Ratzel's political geography. While the lived geography of *pays,* emerging as it did from spontaneous and reciprocal relationships between groups and their natural surroundings, might indeed be described in organic terms, the artificial construction of states could not be (Vidal de la Blache 1896). The French School of human geography stressed the active and creative role of human groups as they fashioned their *genres de vie* within diverse milieus (Vidal de la Blache 1922; Berdoulay 1982; Buttimer 1971). Organic analogies abound in the writings of Vidal and his disciples, but they were understood in their metaphorical rather than literal sense (Berdoulay 1982). Camille Vallaux and Jean Brunhes were both skeptical about the metaphor itself, Vallaux taking pains to show the essential differences between biological organisms and the "surface of the earth" and even counseling that the term be dropped altogether (Vallaux 1925). Vidal, however, insisted on its value as *poesis,* its power to evoke a sense of the whole, as descriptive and heuristic horizon for curiosity about the associations between human groups and their milieus (Buttimer 1971, 1978). More literal and more analytically oriented uses of the term *organism,* for example, in Durkheim's sociological theories, were virtually taboo for the Vidalians. On the extended controversy between Ratzel's anthropogeography and the Durkheimians' social morphology, they would certainly find themselves closer in spirit to Ratzel, while denouncing those aspects of his work which implied environmental determinism (Vidal de la Blache 1913; Berdoulay 1978; Buttimer 1971).

There was a strongly organicist flavor in the "home area studies" (*Heimatkunde*), which became a popular genre throughout northern Europe (Herbertson 1905; E. Kant 1934). It held special appeal in the geographically diversified outer regions from the Celtic fringe to the Balkans, from Catalonia to Baltic lands. Mythopoeically, one could associate organicism with Narcissus among scholars who taught and wrote about civilization and landscape and who advocated sensitivity to ecology and the sense of place. For many, too, such reflections on society and milieu served a Phoenix function, as they strove to affirm the identity and niche of human groups within their natural surroundings and to offer geographical arguments for social justice and political autonomy for oppressed peoples. In some cases, the *Zeitgeist* of imperialism and colonial expansion, the Faustian element, lost touch with its Phoenix roots. It is in those efforts to exploit the metaphor in justifying such political and ideological endeavors that one finds the fundamental explanation for its dismemberment and demise soon after World War I.

Eadem ratio extremorum (extremes derive from the same root). While organism might well be used as metaphor to describe the character of small-scale medieval communities in place, the transposition of this metaphor to society in general in the era of enlightened despotism evoked fear and revulsion (figs. 38, 39).

Organism and the Frontier

In North America, however, on the advancing westward frontier of European migration, organicism appeared to hold unequivocal appeal. From its foundation as a discipline in America, geography echoed this metaphorical style and world-view. In 1751 Benjamin Franklin, an ardent devotee of French scientific humanism, had declared: "A nation well regulated is like a polypus: take away a limb, its place is soon supplied; cut it in two, and each deficient part shall speedily grow out of the part remaining. Thus, if you have room and subsistence enough, . . . you may, of one, make ten nations, equally populous and powerful, or rather, increase the nation tenfold in strength" (cited in Hacker 1947, 113).

For the migrant, the transatlantic voyage symbolized a Phoenix adven-

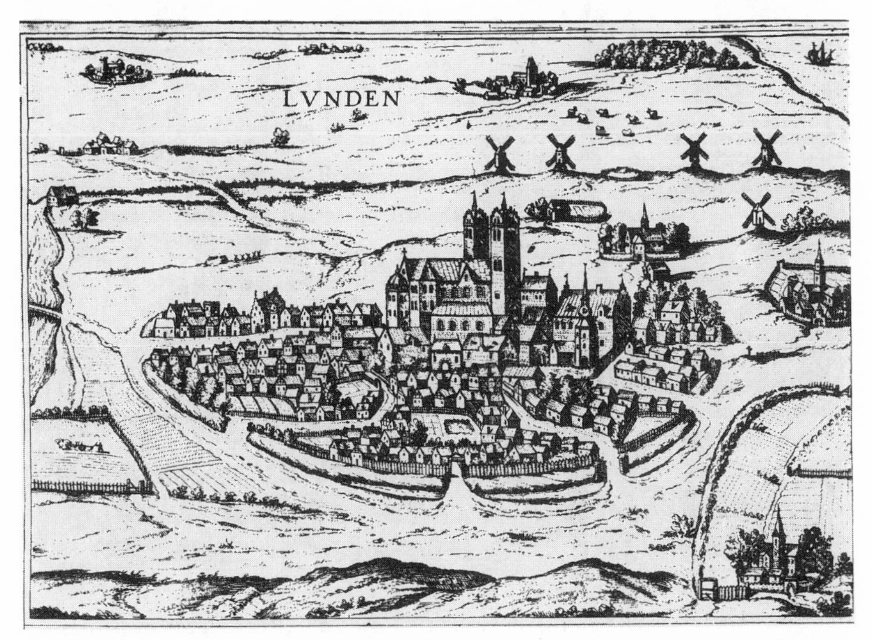

Fig 38. The medieval community of Lund. (Photo courtesy of Lund University Library.)

Fig. 39. "Society as Organism," in Hobbes's *Leviathan*. (Reproduced by permission of the British Library.)

ture; for the would-be designers of this new society, the challenge had already been construed as Faustian. By the mid-nineteenth century, the interaction of Phoenix and Faust in this land of adventure was played out in ways altogether different from those of continental Europe. The drama of the North American frontier attempted to maintain its Phoenix character; the emancipatory saga of a people committed to finding new ways, its story recounted in thoroughly organicist terms: "The United States lies like a huge page in the history of society," Frederick Jackson Turner wrote; "line by line as we read this continental page from West to East we find the record of social evolution" (1894; see also Livingstone and Harrison 1981). "The struggle of the primitive American community to bring the wilderness under cultivation made the practical-minded engineer, who had no time for leisure and no interest in play or foreign culture, the ideal character for everyone to imitate" (Moore 1957, 238).

Arnold Guyot (1807–84), a Swiss immigrant and a disciple of Ritter and Hegel, was the first official geographer in America; he waxed eloquently on the westward "geographical march of history." Nathaniel Shaler, mentor of William Morris Davis at Harvard, was less sanguine in his assessment of the frontier. Having studied the suitability of various continents for the development of civilization, he held the view that Scandinavia, Greece, the British Isles, Spain, and Switzerland were able to nurture great races, whereas America was just not suited for such (Shaler 1894; Koelsch 1979; Livingstone 1982). George Perkins Marsh warned about the inevitable consequences for nature if Americans were blindly to follow a peculiar myth of progress (Marsh 1864). Later, Ellen Churchill Semple narrated the history of America in terms of physical constraints and opportunities; in her Mediterranean study, she illustrated the mutual interpenetration of nature and history in classical culture (Semple 1911, 1931). The strongly environmentalist tone of early American geography could possibly be explained in terms of the personal careers of strong individuals who maintained their ties to European masters.[6] Much of it could also be explained in the context of the rich and varying physiography of that continent itself and its diversity of cultural styles evident at the settlement frontier as it edged its way westward (Phillips 1970). To the extent that organicist approaches could support a progressive evolutionary myth about America itself, it held appeal. But on the horizons of science and the analytical frontier, the challenge of Darwinism caused divisions and controversy (Wright 1961; Stoddart 1966; Livingstone and Harrison 1981).

Origin of Species was reviewed very positively in the *American Journal of Sciences and Arts* in 1869, but the response from geographers was cool. All three major personalities of the day, Louis Agassiz (1808–82), Arnold Guyot

(1807–84), and Matthew Fontaine Maury (1806–73), denounced it and felt that it had nothing to offer geography. Their rejections were often articulated on theological grounds.[7] George Perkins Marsh was evidently not much impressed by theories of natural selection, and Shaler was dubious (Stoddart 1981; Koelsch 1979).[8] Very few of the founding fathers of American geography were trained in biology, botany, or zoology; most had studied geology or physiography. John Wesley Powell (1834–1902), a geologist by training and an eager student of ethnology, was one of the first to express enthusiasm for evolutionary ideas. He was careful, however, to note how inappropriate it would be to apply the laws of biotic evolution to humankind.

William Morris Davis (1859–1936) was an ardent advocate of organicist approaches to geography. "The spirit of evolution," he wrote in 1909, "has been breathed by students of the generation now mature all through their growing years, and its application in all lines of study is demanded" (1906, 6). The famous "cycle of erosion" was for him analogous to the life cycle, and the theory of evolution afforded a route for geography to seek causal explanations of landscape. "The most notable geographic phrase in the English language in the twentieth century," Charles R. Dryer remarked in 1920, "was the Davisian formula 'physiographic control and organic response'" (8). The critical energizing motif, it seemed, was the Promethean myth of progress, defined as increasing control of nature and environment.

Friedrich Ratzel visited the United States in 1874–75 and did not hesitate to express his shock at the apparently profligate waste of resources by this pioneering society. The first volume of his *Anthropogeographie* (1882) was translated by Semple, but ideas such as "the State as organism" found little welcome among Americans. There were some discussions of Halford Mackinder's Heartland Thesis and Alfred Mahan's theory of Sea Power, but generally speaking, there was little overt concern expressed about the issues of geopolitics and *Lebensraum* which stirred emotion among their European colleagues at the turn of the century. There was so much to explore at home, on this continent of allegedly unlimited resources; so many reasons to celebrate the onward march of progress and to justify this in environmentalist terms. Russell Smith's *North America,* one of the most popular textbooks of its day, provided evidence that America had the most favored location of any civilization: its range of temperature and moisture, its topography and resources, provided it with every conceivable advantage (1925). Mark Jefferson celebrated the frontier of communication in his famous essay on "the civilizing rails," mapping out what he defined as the *oecoumene,* namely, land within ten miles of a railroad (1928). "For our purposes," Ellsworth Huntington wrote in the 1920s, "we may define progress as increasing ability to dominate the forces of nature" (1926, 136). Further, "One of the most distinctive features of the American School of Geography is its recognition that the indirect effects of physical environment are at least as potent as the direct effects, and probably more so. . . . Another distinctive feature is its recognition of the importance of health and energy as primary factors in determining the rate of human progress" (vi).

Ellsworth Huntington's geography was widely read by the public at large but was bitterly criticized by his professional colleagues, whose intellectual orientation was shifting away from organicist prose with its ideological connotations. Issues such as ethnicity, race, criminality, and disease may have been deemed more appropriate domains of inquiry for sociologists or for immigration authorities. Besides, hypotheses about environmental influences on human health and productivity were clearly unwelcome in a Promethean clime. The geographer should invest energy in macrospatial patterns—cities, transport lines, industry—demonstrating that branches of "human" geographic inquiry could be pursued just as scientifically as the "physical." Davis, who dared to continue with an integrated approach, ended up by offending colleagues on both sides, geomorphologists accusing him of inadequate empirical justification for his theories, and human geographers, fleeing from the notion of organism, accusing him of mysticism (Dickinson 1969, 119–22). Huntington, Davis, and Semple, scholars well attuned to concurrent developments in Europe, stand out as major articulators of organicism in early twentieth-century American geography.

Any tracings of organicism in American geographical thought during the twentieth century, however, inevitably lead to the Midwest and especially to the University of Chicago in the twenties and early thirties. Sociologists such as Robert E. Park and Roderick D. McKenzie and their associates enthusiastically embraced ideas such as "society as organism," freely playing with metaphors drawn from plant ecology and ecosystems. Harlan Barrows's 1922 presidential address to the Association of American Geographers sought to weld the fields of human geography and human ecology. It was a homily that perhaps reflected the orientation of his colleagues in sociology and anthropology at the University of Chicago more than the emerging formistic perspective of his fellow geographers across the nation. It certainly echoed George Perkins Marsh's concern for conservation of America's natural resources (Marsh 1864), and did spark environmentalist concern among students—for example, Gilbert White (G. White 1974; G65). Meanwhile, on the research frontier of American anthropology, the leading figures, Franz Boas, Robert H. Lowie, and Alfred L. Kroeber, found evidence of organicity in the relations between cultural groups and their natural milieus. During the first quarter of the twentieth century,

American social scientists conducted debates remarkably analogous to those aired between Ratzel and Durkheim in latter nineteenth-century Europe (Buttimer 1971; Berdoulay 1978). Similar antinomies were cited: "organic" versus "superorganic," "biotic" versus "cultural" (Theodorson 1958; Duncan 1980). The mystique of "natural communities" in sociology, like that of "natural regions" in geography, both inspired by the Haeckelian notion of "ecosystem," illustrates the tenacity of organism as root metaphor.

Programmatic statements about the nature of geography during the twenties were few, and they tended to cite European authorities rather than scrutinize the epistemological grounds for the discipline's cognitive claims. This practice often obfuscated important analytical distinctions, as Stoddart and Mikesell have demonstrated with respect to Darwin and Ratzel (Stoddart 1966; Mikesell 1968). Questions of time, process, organicity, and contingency were rarely explored. As might be expected, there were other factors involved, such as translation difficulties among more than three vernacular languages; strong personalities; institutions of education, commerce, and research; and shadows of war and ethnic stereotypes (Dickinson 1969). Such confusion prompted many Americans to dismiss appeals to authority altogether and to get on instead with the business of developing methods and practices that could work. There were appeals about the unity of geography, programmatic statements about the relation of inductive and deductive reasoning, and persistent arguments about the importance of field analysis. Organism as root metaphor was, however, to lose its appeal.

In the wake of World War I, notions such as environmental determinism, "control," "influence," "relationships between organic and inorganic," and, most of all, the idea of state totalitarianism defended with terms such as race and territory, became distasteful to most American geographers. In a sense, their rejection of organicism might be construed as a rejection of European masters, of deductive logic and rational models, or rhetoric that could not be tested operationally. The Anglo-American preference for empiricism, coupled with an American penchant for pragmatic problem solving, undermined the credibility of organicism among American geographers. Even compelling pleas for an ecological orientation within geography, such as those by Huntington, Davis, and Barrows, bore within them the seeds of alternative metaphors. One direction was obviously the analysis of systems that eventually converged with mechanism. Another was to deal pragmatically with each event as it occurred, the contextualist route followed only by a few. The overwhelming choice made by American geographers during the twenties and early thirties was the mosaic, which means that these individuals adopted (explicitly or implicitly) the Kantian definition of geography as chorology. (See Chapter 3.)

Organicism did not immediately disappear from the horizon. Historians of science and researchers on human creativity have frequently approached their subject in organicist ways. But the notion of organism certainly held stronger appeal for rural sociologists (Zimmermann and Frampton 1935), a few biologists such as Barbara McClintock (Keller 1983), and workers' communities and the New Deal "back-to-the-land" movements (Nisbet 1953; Bookchin 1974) than it did for geographers of the interwar period. At the American Geographical Society in New York, Richard U. Light continued in the spirit of Jacques May by promoting geographical study of health and disease (Light 1944). For the majority, however, metaphors of mosaic and mechanism offered ample scope. Not until the dawning of environmental consciousness in the late sixties, the eventual assertion of existential humanism, and the rediscovery of the classical notion of *genre de vie* did organicism again win an audience among American geographers. In the late eighties, as challenges of the global environment flashed on the research horizon (WCED 1987; Mungall and McLaren 1990), geographers were challenged to reconsider their long-established metaphors, again to explore the cosmos, within and without.

Organism, *Poesis,* and Human Identity

The primary contribution of organism as root metaphor for geography was its capacity to evoke curiosity, emotion, and the will to discover the threads of cohesion in landscape and life. This *poesis* served well in establishing the identity of the discipline itself, offering horizons for research endeavor which included both human and physical branches of inquiry. It also opened doors for cross-disciplinary exchanges of insight, particularly with history and biology, and eventually with anthropology, sociology, and medicine. In terms of *paideia,* organicist geography exercised tremendously evocative power. Here one must recall that many of the early teachers of geography delivered their lessons orally in field excursions, laboratory exercises, and supervised local area studies. The aims of geographical education included not only promoting understanding and a synthetic grasp of life and landscape but also fostering patriotism and sensitivity to natural phenomena. The *Heimatkunde* of German-speaking lands, the *pays* studies of French-speaking parts, the local area research in Britain, and *Hembygdsforskning* in Scandinavia constituted essential components of *Bildung*.

These qualities of multisensory appeal and synthetic approach to learning, however, made organism less amenable to *logos* and *ergon,* especially when these two lines of professional endeavor submitted themselves progressively to the norms and strictures of scientific rationality defined in positivist terms and to the concurrent imperatives of political expediency. Where *logos* was concerned, organicism could lend itself to transposition into mechanistic terms, thereby gaining in analytical rigor and nomothetic potential, but losing the sense of unity and cohesion in its overall image of reality. Organic ideas presented inspiring *poesis* for applied endeavor, for example, in garden cities, regional development plans, and scenarios of empire. Relying as it did on emotional, aesthetic, and moral appeal, however, such rhetoric often led to the abandonment of reason and the legitimization of power hierarchies which showed little concern about the overall dynamics of the cosmos.

None of these practices can be understood until one considers the milieus in which they were expressed at different moments of Western history. Among the human interests that geography addressed, it was certainly *identity* that organicism elucidated best. Geography could reveal the extent to which human identity (local, regional, cultural, national) was intimately related to landscape and resources (*niche*). Herein lay the dilemma: the same rhetoric and discipline that had led young students to a knowledge and affection for their home areas could later be exploited for chauvinistic and totalitarian politics. The idea of world as organic whole has played a Phoenix role in creativity within cultural, political, and intellectual life at different moments in Western history. Once it was exploited for the power-political goals of expansionary regimes, or its rhetoric used to promote the special interests of any one nexus within the organic whole of the earth's anthroposphere, its integrity was compromised. Geographers of formerly imperial nations now criticize organicism as a legitimating rhetoric for imperialism, racism, and a host of other evils (Eisel 1976; Schultz 1980; Harvey G34, 1984). For the scholars of the so-called peripheral lands and of postcolonial and politically marginal corners of the earth, organism had a very different kind of metaphorical meaning. In their actual landscapes, as in the myths and symbols of their own civilizations, many have come to discover an authentic sense of identity, order, niche, and horizon (Bonnemaison 1985; Chapman 1985; Karjalainen 1986). With the growing momentum of geopolitical change in the latter twentieth century, lessons from the past might indeed merit some reflection.

Within the world of science itself, the 1980s also witnessed the emergence of a new consciousness about the interrelatedness of living reality. In his vision of self-generating systems, Nobel laureate Ilya Prigogine announced a fresh alliance of science and humanism and a new integration of knowledge with implications for all branches of research: "For the ancients, nature was a source of wisdom. Medieval nature spoke of God. In modern times nature has become so silent that Kant considered that science and wisdom, science and truth, ought to be completely separated. We have been living with this dichotomy for the past two centuries. It is time for it to come to an end" (Prigogine and Stengers 1984, 98).

James Lovelock's *Gaia* (1979) was surely an appeal for organism as metaphor for terrestrial reality:

> Ancient belief and modern knowledge have fused emotionally in the awe with which astronauts with their own eyes and we by indirect vision have seen the Earth revealed in all its shining beauty against the deep darkness of space. Yet this feeling, however strong, does not prove that Mother Earth lives. Like a religious belief, it is scientifically untestable and therefore incapable in its own context of further rationalization. (vii)

Chapter Six

World as Arena of Events

They had scarcely set foot in the city, mourning the death of their benefactor, when they felt the earth tremble beneath them. The sea boiled up in the harbor and smashed the vessels lying at anchor. Whirlwinds of flame and ashes covered the streets and squares, houses collapsed, roofs were thrown onto foundations and the foundations crumbled; thirty thousand inhabitants of all ages and both sexes were crushed beneath the ruins.

The sailor whistled, swore and said, "I'll get something out of this."

"What can be the sufficient reason for this phenomenon?" said Pangloss.

"This is the end of the world!" cried Candide.

The sailor immediately rushed into the midst of the wreckage, braved death to find money, found some, took it with him, got drunk and, after sobering up a little, bought the favors of the first willing girl he met in the ruins of the destroyed houses, amid the dead and dying. But Pangloss pulled him by the sleeve and said to him, "You're behaving badly, my friend: you're not respecting universal reason, you've chosen a bad time for this."

"By the blood of Christ! I'm a sailor and I was born in Batavia: I've walked on the crucifix four times during four stays in Japan—you've come to the right man with your universal reason!"

Candide had been wounded by several splinters of stone. He was lying in the street, covered with rubble. He said to Pangloss, "Alas! Get me some wine and oil: I'm dying."

"This earthquake is nothing new," replied Pangloss. "The town of Lima in America felt the same shocks last year. Same causes, same effects; there is surely a vein of sulphur running underground from Lima to Lisbon."

"Nothing is more likely," said Candide, "but in the name of God, bring me some oil and wine!"

"What do you mean, likely?" retorted the philosopher. "I maintain that the fact is demonstrated."

Candide lost consciousness, and Pangloss brought him a little water from a nearby fountain.

—Voltaire [1759] 1959, 29–30

IN A GENRE later echoed in Lewis Carroll's *Alice in Wonderland* and Saint-Exupéry's *Little Prince,* Voltaire's *Candide* satirized the conventional beliefs and lifeways of his eighteenth-century world. Stories about travel through exotic places and events, real or imaginary, rank high among the *chef d'oeuvres* of Western literature. "Journey" itself is a metaphor provoking insight into the human experience of world, from individual lives through those of nations, societies, and empires. Geographers have journeyed far and near, often pondering questions of language, method, and style in which to render the foreign and exotic comprehensible. Herodotus and later explorers recounted the saga of events and places encountered on their voyages. Hecataeus and Ptolemy, on the other hand, proposed a more general language and grid system within which to deliver geographic information. Since earliest times, geography has had many a Ptolemy and many a Herodotus; some generalizing about landscapes, patterns, and processes, others focusing on the uniqueness of particular places and events.

To picture the world as a stage on which spontaneous and possibly unique events occur is the characteristic trait of contextualism, fourth among Pepper's world hypotheses. Events, in all their complexity, possible uniqueness, and contingency, capture attention here; one looks for holistic understanding of particular events rather than ways of fitting them into some a priori schema of form, process, or organic whole. *Arena* seems to be the most appropriate metaphor to capture contextualist approaches in geography. This is an elusive world-view, one whose interpretations of reality at times converge with others. In its synthetic approach to reality, it bears resemblances to organism, but it resists integration into universal wholes. Like mosaic, arena projects a dispersed world-view; but, unlike all others, it distrusts fixed categories and finds analysis potentially distorting. There is a "catch-me-if-you-can" motif running throughout, and any attempt to trace linear sequences in the genesis of this root metaphor would be inappropriate for this genre. If mechanism might be associated with revolution, formism with the status quo, and organicism with tensions between conservative and radical views, then contextualism would most characteristically be associated with liberal stances on political life (H. White 1973).

Pepper identified contextualism with American pragmatism, a philosophy imbued with the energy and sometimes naive optimism of a frontier society. Genial toward any and all potential explanations of reality, pragmatism sought to transcend the competing claims of inherited approaches by appealing to both "tough-minded" empiricist and "tender-minded" rationalist (W. James [1907] 1955, 22). William James equated it with humanism, and indeed there was an emancipatory tone in its irreverence for tradition and its quest for practically useful and workable solutions to problems: "Pragmatism unstiffens all our theories, limbers them up and sets each one at work. Being nothing essentially new, it harmonizes with many ancient philosophic tendencies. It agrees with nominalism, for instance, in always appealing to particulars; with utilitarianism in emphasizing practical aspects; with positivism in its disdain for verbal solutions, useless questions and metaphysical abstractions" (James 1955, 46).

Pepper's description of contextualism was articulated within the frame of reference of the twentieth century's first four decades. For illustrations he cited works authored by John Dewey, George Herbert Mead, Alfred Schütz, and members of the American school of symbolic interactionism. Unprecedented events and profound transformations of context have transpired since then. Fifty years later, arena evokes a kaleidoscope of images, from the visions of theoretical certainty projected by enthusiasts of chaos dynamics and social engineers at one end of a spectrum, to the solipsism and *reductio ad absurdum* of postmodernism and critical musings about language, knowledge, and power at the other. Following the spirit of Pepper's formulation, however, some classical foundations and periodic expressions of this contextualist approach throughout Western history are traceable.

From Sophistry to Pragmatism

"Man is the measure of all things," Protagoras (ca. 490–after 421 B.C.) argued in a treatise, *Truth,* "of what it is, that it is, and of what is not, that it is not." Skeptical of philosophy and all a priori theories about the nature of the universe, Sophists denied the possibilities of objective truth. Reality is always filtered through the lenses of the perceiver, it was argued, one's perceptual abilities being the measure of reality. Why, then, bother with philosophy and the vain pursuit of truth? Better to learn prudence and efficiency in making a living and in assuming a useful role as citizen of the state. The Sophists evidently were charismatic teachers, their doctrines especially appealing to youth, so that initially at least they were regarded as subversive by the elders.

Socrates shared the Sophists' skepticism about philosophy and applauded their educational aims. On monetary reimbursement by the state or sponsor, however, he conscientiously objected. Nor was the training of professionals alone deemed an adequate goal for the teacher; the integral development of the whole person, of intellectual as well as moral virtues, not just the training of experts, should be the goal. Though acclaimed by subsequent generations as the philosopher par excellence, Socrates never set absolute truth as the goal of study; rather, it was to reach an understanding of the limitations of human understanding.

In Periclean Athens, the Sophists enjoyed popular appeal, as they proclaimed the value of education and rhetoric for practical life and rarely missed an opportunity to disparage the philosophers and their speculations about physics and metaphysics. By the time of Plato's *Republic,* however, the tables were turned; even Isocrates, one of the most famous Sophists, once claimed the title of philosopher. Now it was Plato's turn to retort in anything but flattering tones: "Herein, resembling some tinker, bald-pated and short of stature, who, having made money, knocks off his chains, goes to the bath, buys a new suit, and then takes advantage of the poverty and desolation of his master's daughter to urge upon her his odious addresses" (*Republic,* vi.495E).

Already in the fifth century B.C., features of the world-view that Pepper labeled contextualism can be found. First, there is a skepticism about philosophy, universal theories, and discourse about truth. The Heraclitean theory that the universe is always in a state of flux, for example, was the example that Protagoras used as evidence of the futility of intellectual pursuits. Second, there was the emphasis on practical relevance to everyday life: successful performance in rhetoric, business, or politics offered a sounder criterion of value than logic or morality. Rhetorical elegance in winning an argument was deemed more important than demonstrating the epistemological grounds of one's case. Third, there were hints of what would now be called "hermeneutic" awareness, namely, that conceptions of reality were filtered through the lenses of individual perceivers. From such awareness, however, two very different implications were drawn: the Protagorean choice of expediency in one's approach to life and the Socratic choice of critical reflection, self-understanding, and ultimately awareness of ignorance.

Within the heart of the Pax Romana, as within those of later empires, a critique of conventional ways was expressed. Cato, Pliny, and Juvenal satirized the lifeways of later imperial Rome, as Archimedes had done in Hellenic Greece. Around the turn of the first century A.D., Juvenal launched his *Book of Satires,* "the motley tale of all man's life: his vows, his fears, his anger, his pleasures, joys and journeying to and fro." In strident verse, he exposed the vanity and ostentatiousness, pride and prejudice, fads and injustices, *panes et circenses,* of an empire on the wane (Carcopino 1960). From within many a subsequent regime in the West, satirical voices have echoed the Juvenal style, from Montaigne to Pope, Samuel Johnson to Oscar Wilde, Balzac to Strindberg. In these masterpieces, generally acclaimed for their artistic style and poetic qualities, the intellectual historian could find evidence of Narcissus, reflecting on the mirrors and masks that characterized social interaction at different times. *Poesis* indeed has often expressed itself in a satirical voice, with irony and skepticism, to expose hypocrisy and Pharisaism in the taken-for-granted, and eventually to clear the ground for alternatives. These texts were rendered publishable, however, because beneath all the satire and critique, there lurked an ideal model for *humanitas,* deviations from which were exposed, and a return to which might have been the primary motivation of the author.

Some, of course, would seek wider horizons within which the theater of everyday events could be interpreted, a quest often shared with organicists. To gain a fuller grasp of events, one needs the largest possible context of space, time, and cultural horizon within which events could be seen unfolding.[1] Still, the contextualist would have avoided, as long as possible, causal explanations of events, emphasizing instead the unpredictable and spontaneous contingencies of human history.[2] Montaigne, who was at pains to point out the fallibility and mutability of commonly held notions of scientific knowledge, wrote:

> Ptolemy, who was a great man, had established the bounds of this world of ours: all the ancient philosophers thought they had the measure of it . . . and behold! in this age of ours there is an infinite extent of *terra firma* discovered, not an island or a particular country, but a part very nearly equal in greatness to that we knew before. The geographers of our times stick not to assure us, that now all is found, all is seen. . . . But the question is whether, if Ptolemy was therein formerly deceived, upon the foundations of his reason, it were not very foolish to trust now in what these later people say: and whether it is not more likely that this great body, which we call the world, is not quite another thing than what we imagine. (Essays, bk. 2, chap. 12)

The legacy of the Sophists in Western literature, rhetoric, politics, and education is not so easily retrievable from printed records. It is characteristically on the periphery of established regimes that critical voices have drawn attention to the here-and-now, to the practical consequences of conventional theories, to the apparent paradoxes of thought and action, rule and practice, in science as well as in political life (O'Neill 1982). The spirit of Protagoras was voiced at twilight moments down through the centuries, sometimes as a dirge for cultural Pharisaism, sometimes as a satire on the encrusted certainties of the parents' generation, sometimes to herald radically fresh alternatives for human ways of thought and being. It would find a fresh lease on life among American pragmatists in the late nineteenth and early twentieth centuries.

"Shall we never, never get rid of the Past?" Holgrave moaned in *The*

House of the Seven Gables, "It lies on the present like a giant's dead body! We laugh at dead men's jokes, cry at dead men's pathos" (Lewis 1955, 18–19). Thoreau regarded England as "an old gentleman who is travelling around with a great deal of baggage, trumpery which has accumulated from long housekeeping, which he has not the courage to burn" (Lewis 1955, 21–22). Nathaniel Hawthorne's "The Earth's Holocaust" heralded the burning of all European aristocratic trappings, institutions, literature, and philosophy. America was to be the land where one could be born again in a rustic paradise away from cities already infested with demons from urban Europe (Hawthorne [1844] 1972, 887–906); born again to "true" religion, and, in John Winthrop's words, "raising a rampart against the kingdom of Anti-Christ which the Jesuits are in process of establishing in these regions" (Eliade 1969, 95).

By the latter years of the nineteenth century, however, the dream of an earthly paradise based on a return to nature and Puritan faith had become transposed to a Faustian myth of secular progress. A more instrumental attitude also developed—at least at Harvard and Chicago—in the movement known as pragmatism. It assumed a radically empiricist attitude toward truth and being, turning its back on abstractions, a priori principles, and closed systems, and facing toward concreteness, adequacy, and action. William James summed it up:

> "The true," to put it very briefly, is only the expedient in the way of our thinking, just as "the right" is only the expedient in the way of our behaving. Expedient in almost any fashion; and expedient in the long run and on the whole of course; for what meets expediently all the experience in sight won't necessarily meet all further experiences equally satisfactorily. Experience, as we know, has ways of boiling over, and making us correct our present formulas . . .
>
> We have to live today by what truth we can get today and be ready tomorrow to call it falsehood. Ptolemaic astronomy, Euclidean space, Aristotelian logic, scholastic metaphysics were expedient for centuries, but human experience has boiled over those limits, and we now call these things only relatively true, or true within those borders of experience. "Absolutely" they are false; for we know that those limits were casual, and might have been transcended by past theorists just as they are by present thinkers. (1955, 145)

Charles Sanders Peirce (1839–1914), himself trained in positivism, launched the term *pragmatism* in the latter years of the nineteenth century. Philosopher and physicist, he worked for thirty years at the United States Coast and Geodetic Survey (Wiener 1949). Attacking the belief in "ultimate foundations" of both rationalist and empiricist schools, he pointed out, "There can be no knowledge which is not mediated by prior knowledge"

(Peirce 1931–35, 5: 259). "The unanalyzable, the inexplicable, the unintellectual runs in a continuous stream through our lives" (5: 289). Reflecting critically on the positivist enterprise in science, he tried to shift the focus of concern away from methodology and logical structures of scientific theories to the logic of inquiry. Anticipating more general awareness about the social construction of scientific ideas, he claimed that the institutionalization of particular modes of inquiry determined beliefs about cognitive validity. Why not, then, examine those ways in which knowledge is constituted? Science itself was a "mode of life" (7: 54).[3] If reality is defined as the totality of all possible true statements, and if these statements are symbolic representations, then why should the structure of reality *not* be elucidated in relation to the structure of language?

Contemporaries of realist and idealist orientation were as harsh in their critique of pragmatism and its instrumental approach to truth as Plato had been of Isocrates. Its influence on subsequent developments, especially in sociology, psychology, and landscape analysis, however, has been profound. John Dewey's instrumentalist theory of knowledge left its mark on several generations of educational policy makers in America, and George Herbert Mead's pioneering ideas on the social construction of knowledge have found enthusiastic audiences in geography and anthropology (S. Smith 1984). With its eclectic attitude toward inherited theories, pragmatism aimed at steering a middle ground between realism and idealism. It dismissed traditional distinctions between ways of knowing and ways of being, emphasizing process rather than pattern, as well as the tentative and fallible nature of all theory. It was in this spirit that Pepper described the cognitive claims of contextualism.

An Operational Theory of Truth

A contextualist sees everything in life in terms of the ever-changing present *event*. Categories of analysis are derived from common-sense experience of the event itself—the act within its context. One speaks in verbs—doing, creating—and all events are intrinsically complex, like incidents in the plot of a drama. Disorder and change are regarded as the normal state of affairs. No complete or final analysis of the world is ever possible, and the very idea of unraveling the "elements" and "structures" of reality, as mechanism does, is regarded as absurd. In the contextualist light, the universe is one of contingency and random incidents. One might speculate outward from the *quality* and *texture* of particular events, but ultimately the quality of a given event can only be intuited in holistic terms. Qualities attract one's attention but not one's analytical gaze. Textures might be discernible in details and relations, but these compromise quality. The holistic and synthetic grasp of an event demands attunement to diverse strands and references criss-crossing within the *context* of an event, but it also demands a discernment of "quality" or character of that unique happening. If the account "works," namely, if it yields a plausible explanation of that particular event, it is regarded as true. The theory of truth on which the root metaphor rests is known as the *operational theory of truth*.

Quality refers to the total meaning of an event, a wholeness of understanding that may be discerned in terms of three major categories:

1. The *spread* of an event (its "specious present"): each event defines its own time- and space-reference: its "reach" to past and future. A distinction is made between qualitative (*durée*) and schematic (clock-calendar) time, for instance. In the actual event, the present is the whole texture that directly contributes to the quality of an event: actual time is the forward and backward spread of the quality.
2. *Change:* absolute permanence or immutability is regarded as fictional. What may appear as permanent is interpreted in terms of historical continuities, which are not changeless.
3. *Fusion:* whatever appears simple and unified in experience is actually the result of fusion, for example, flavors or colors—qualities of the details get fused in the quality of the whole. Usually there are degrees of fusion, and the tighter the degree of fusion the greater the unification. Qualities such as "good," "beautiful," and "appropriateness," for example, are the result of fusion. The contextualist would insist that this is no psychological rationalization, but rather a reflection of the active structures of textures. These qualities give meaning to all the textures of experience.

Texture is conceived in terms of strands, context, and references. A texture is actually made up of strands, and it lies within a context. In fact, it is the connections of strands which determine the context, and in turn the context determines the qualities of the strands. No sharp lines may be drawn between texture, strand, and context, hence element analysis is essentially distortive. The only rationale for analysis is pragmatic, namely, to find solutions to specific questions.

(1) *Strands* and (2) *context* interweave in the unfolding of events. Together they are associated with (3) *references,* of which at least three kinds can be identified:

a. *Linear references* have a point of initiation, a transitive direction, and achieve an ending in satisfactions. These references can be seen to have a backward-and-forward activity.

b. *Convergent references* involve several initiations converging upon one satisfaction.
   Both linear and convergent references may encounter *blocking,* which interrupts the direction implied in any set of initiations. Total blockings may lead to novelties and surprises.
c. *Instrumental references* can imply linear references that have been blocked and a secondary reference that neutralizes the blocking. The result is often a complex texture extending far beyond the limits of a given event.

Contextualists argue that in-depth scrutiny of any private event inevitably carries one into a public world. Interpenetration of textures can be observed not only in social events but also in human perceptions. Knowledge is thus entirely relational. One does not know "quality" except in terms of one's perceptions, but one can make inferences about its texture outside of perception. Such relational knowledge of textures consists in those schemata such as maps, symbolic systems, and diagrams, which are based on social experience. They are part of the science of a period and change from one period to another.

As outlined by Pepper, this operational theory of truth has three distinct specifications. There is the "successful working theory," which tests the validity of a proposition in terms of how well a hypothesis works. There is a "verified hypothesis theory" wherein truth is not based on the success of an act but rather in the relation between hypothesis and its eventuality. Like their positivist mentors, pragmatists would not claim that the true hypothesis itself delivers insight into the qualities of nature. A third specification is the "qualitative confirmation" one, which assumes that the meaning of a symbol is found in the quality it leads to, and that the quality of a strand takes up the qualities of its context. The texture of perception shows that the texture of the verifying act must be partly made up of the strands carried into it by the activities of the perceiver. The body of hypotheses possessed by science and philosophy at any moment should thus afford insight into the structure of nature.

Divergent Streams in Europe and America

For the American pragmatist, the world as arena was a world of infinite possibilities, one in which human ingenuity confronted challenges and handled them expediently without wasting energy on metaphysical debates or philosophical reflection. For Europeans, however, such a metaphor probably elicited different associations around the *fin de siècle*. Not only was the truth of yesterday less easy to dismiss, but that of the day was radically challenged. If North American contextualists regarded themselves as being at the dawn of a new era, many of their European colleagues were experiencing the twilight of great empires, soon to be dismembered after 1914. Among those who did not emigrate westward, some renewed faith in Enlightenment (e.g., Weberian) rationality, thereby resuscitating the worldviews of mechanism and organism; others espoused radical empiricism or logical positivism. In France and Germany, as indeed at Harvard, heated debates arose about evolution in life as well as in thought, about free will versus determinism, and about ideological influences on the conduct of science. Geographers wrestled with questions of environmental influences, while philosophers wrestled with debates over biological evolution. Henri Bergson (1859–1941), whose ideas were later to inspire Alfred Schütz and Maurice Merleau-Ponty, was a powerful defender of human freedom. He claimed that freedom was a fact of life which could not only be sensed intuitively but could also be documented empirically, if research in the human sciences could be organically related to that in the life sciences:

> The theory of knowledge and the theory of life seem to us inseparable. A theory of life that is not accompanied by a criticism of knowledge is obliged to accept as it finds them the concepts which the understanding puts at its disposal. It can enclose the facts, willing or not, in preexisting frames which it takes as ultimate. It thus obtains a symbolism which is convenient, perhaps even necessary to positive science, but not a direct vision of its object. On the other hand, a theory of knowledge which does not replace intelligence in the general evolution of life, will teach us neither how the frames of knowledge have been constructed nor how we can enlarge or go beyond them. It is necessary that these two inquiries, theory of knowledge and theory of life, should join one another and by a circular process push each other on indefinitely. (Bergson 1911, xiii)

Theories could thus be regarded as conventions, convenient ways to symbolize reality, reflections of concurrent societal and material circumstances (Poincaré [1914] 1955). The school of philosophy commonly referred to as "conventionalism" shared much in common with American pragmatism, but the fundamental assumptions were quite different. Physicists such as Werner Heisenberg and Niels Bohr pointed to the import of particular instruments on the process of inquiry even in the physical sciences, what Heisenberg called the "inevitable ripple" between observer and object:

> You cannot obtain any knowledge about an object while leaving it strictly isolated. The theory goes on to assert that this disturbance is nei-

> ther irrelevant nor completely surveyable. Thus after any number of painstaking observations the object is left in a state of which some features (the last observed) are not known, or not accurately known. This state of affairs is offered as an explanation why no complete, gapless description of any physical object is ever possible. (Schrödinger 1967, 135)

Among pioneering members of the Vienna Circle, however, there was a reluctance to dismiss concern about the ontological questions facing science (and politics) of the early twentieth century. Europeans seemed loath to entrust questions of truth and being to the shifting sands of epistemological relativism such as those espoused by American colleagues. "Truth lives, in fact, on a credit system," William James wrote in 1907:

> Our thoughts and beliefs "pass," so long as nothing challenges them, just as bank-notes pass as long as nobody refuses them. But this all points to direct face-to-face verifications somewhere, without which the fabric of truth collapses like a financial system with no cash basis whatever. You accept my verification of one thing, I yours of another. We trade on each other's truth. But beliefs verified concretely by somebody are the posts of the whole superstructure. (143)

North American pragmatists might thus have blithely laid stakes on the operational workability of truth propositions. It was perhaps this reluctance about conflating issues of thought and being that prompted James to deem their work as being close to humanism, some of them missing it only by a "hair's breadth":

> The notion of a *first* in the shape of a most chaotic pure experience which sets us questions, of a *second* in the way of fundamental categories, long ago wrought into the structure of our consciousness and practically irreversible, which define the general frame within which answers must fall, and of a *third* which gives us the detail of the answers in the shapes most congruous with all our present needs . . . is, as I take it, the essence of the humanistic conception. . . . Whether the Other, the universal that, has itself any definite inner structure, or whether, if it have any, the structure resembles any of our predicated *whats,* this is the question which humanism leaves untouched. It insists that reality is an accumulation of our own intellectual inventions, and the struggle for "truth" in our progressive dealings with it is always a struggle to work in new nouns and adjectives while altering as little as possible the old. . . . Bergson in France, and his disciples, Wilbois the physicist and Leroy, are thoroughgoing humanists in the sense defined. Professor Milhaud also appears to be one; and the great Poincaré misses it by only the breadth of a hair. In Germany the name of Simmel offers itself as that of a humanist of the most radical sort. Mach and his school, and Hertz and Ostwald must be classed as humanist. (236)

That "hairbreadth" distinction between Poincaré's approach to science and William James's definition of humanism reflects the contrasting approaches to questions of being (ontology) which characterized contextualist writings on either side of the Atlantic.[4] Conventionalism did apparently find a strong echo in the writings of the Vidalian School of geography (Berdoulay 1978). Few schools could rival this one as trailblazer for contextually sensitive accounts in historical geography, as exemplified in Fernand Braudel's monumental work on the Mediterranean, the studies published in *Annales, Economies, Sociétés, Civilisations,* and those inspired by this approach (Rochefort 1961). During the latter years of the nineteenth and the early decades of the twentieth centuries, however, contextualist thought in geography followed very different courses on the two sides of the Atlantic.

The contexts of sponsorship and audience at the time of geography's foundation as an academic discipline were also quite different. In Europe as well as in America the dominant root metaphors were organism and mosaic. Early attempts to explore contextualist approaches in European schools tended to be eventually articulated in the language and prose style of other metaphors. J. G. Granö's *Reine Geographie* and Jules Sion's explorations into regional *mentalités* led to richer insights into conventional curiosities about form, for example, the nature and shape of perceptual regions or *pays* (Granö 1929; Sion 1909). Jean Brunhes explored psychological elements in geography, and Pierre Deffontainnes noted the cultural diversity in human perceptions and uses of nature. When these insights were eventually incorporated into texts, however, they were presented as elements for systematic generalizations about *genres de vie* (Brunhes 1902; Deffontainnes 1933, 1948). Alfred Rühl's poetic observations on historical and cultural variations in the values attached to natural resources were similarly presented as grounds for alternative approaches to economic geography (1938). Whatever contextualism's potentially poetic appeal, the case for such an approach very often required a sense of a problem to be solved, be it the definition of regional boundaries, policies on land use, or the conservation of natural resources. And the contexts of problem solving in Europe differed radically from those in America.

Leading European geographers expressed misgivings about the discipline's explicit involvement in practical concerns. Despite the official involvement of geographers in the colonial adventures of major nations, there were recurrent pleas expressed by members of geographical societies about geography's status as an academic pursuit and a science. But pragmatic goals were never far from the horizon of pioneering geographers in North America. In 1859, the year when Ritter and Humboldt died, Dr. J. P.

Thompson, vice-president of the American Geographical Society, claimed: "This science of geography, once regarded as a mere matter of dry but necessary information, is now seen to have vital relations to Man in his physical, mental, social, historical, and moral development" (Thompson 1859, 98–107). On both sides of the Atlantic, too, the sirens of geographical exploration lured recruits, and sponsorship was not lacking. Hence, Thompson argued, "Every well-planned geographical expedition repays its cost to commerce a thousand fold. . . . Commerce must ever adjust itself to the great laws of climate and of physical distribution, which geography ascertains and reports for her guidance" (cited in Wright 1952, 44–45).

It is, however, not easy to trace the influence of pragmatism as a philosophy on the practice of geography in America (Wright [1947] 1966, 124–39). Nathaniel Shaler, friend of William James, regarded the earth as an intensely "humanizable" arena, an open universe in which man's creative intelligence could solve all problems (1894, 1904; see also Koelsch 1979; Livingstone 1982). He regarded the frontier as an unfolding saga of human intelligence and skill, advocating detailed geographical study on the potential resources of that Cordilleran region which had come to be regarded as the "Great American Desert":

> The American desert of our older geographies was pictured as the most inhospitable realm, fit to be compared in sterility with the arid wilderness of Asia and Africa. In part, the impression it made upon the early explorers was due to the fact that they went forth into its fields from the densely forested and superabundantly watered district of the eastern part of the continent. . . . In a word, the name of desert, which was applied to the district, is to a great extent a misnomer; it might be better termed "the arid region," or better still "the country of scanty rain." (Shaler 1894, 1: 145–46)

Shaler's convictions about the superiority of Anglo-Saxon culture and his leanings toward biological explanations of cultural traits, however, scarcely allowed scope for the relativist attitudes associated with an operational theory of truth. He never developed a major school, though his disciple, William Morris Davis, became a leading light within the American profession. Shaler's gift lay largely in *poesis* and *paideia:* students flocked to hear and read his ideas. Later generations would be inspired to undertake the *logos* and *ergon,* for example, in W. L. Thomas's *Man's Role in Changing the Face of the Earth* (1956).

In this pragmatically oriented society, however, geography would claim status as a science; and, for the first two generations, it tended to follow European (German and British rather than French) precedents (Blouet 1981). In the historical writings of Ralph Brown and John Kirtland Wright, one finds elements of this root metaphor of world as arena. Wright, in fact, is perhaps the American geographer who best illustrates the multifaceted range of possibilities afforded by this world-view and its implications for the conduct and substantive foci of geographical research:

> Whole broad domains of science are cultivated not for the immediate purpose of formulating general laws but in order to understand specific conditions and processes. This is especially true of geology and geography, where the first objective is to explain the origins, nature, and relationships of *particular* land forms, rock formations, types of settlement, routes of trade, and what not, as they exist in *particular* regions. If the scientific merits of research are judged formally according to the degree to which they succeed in stating general laws, rather than according to the quality of the work devoted to such research, a large part of our two sciences of geology and geography would be denied scientific merit. (1966, 61. Emphasis mine)

Wright's style was rich in metaphor, especially when he wrote on the history of geographical thought. Changing currents of thought and *Zeitgeisten* in the American tradition, for example, were described in climatic terms:

> Indeed, during Puritan times . . . American geographical understanding as manifested in print (especially in New England) absorbed piety from the surrounding intellectual atmosphere as a towel does moisture from a down-East fog. Since then there has been a gradual secular change in the intellectual climate, reflected in a progressive if spasmodic decrease in the "humid" components of openly expressed piety in scholarship of other than a specifically theological nature. (1966, 253)

He then described the "theological drought" of the post–Civil War period and the growing "mathematical humidity" of the twentieth century:

> Mathematical humidity and theological aridity may increase simultaneously. Thus while the American air has become desiccated of pietistic humidity since Puritan times it has also become ever more heavily charged with the vapors of mathematics and statistics, under whose fructifying influence the young growths have flourished and changed into mighty rain forests and tangled jungles. (1966, 287)

A pluralistic attitude toward modes of knowing and being, a keen sense of history, and the gift of evocative literary style were among the qualities most appreciated by humanist followers of Wright. One finds this strongly contextualist flavor in the writings of Meinig, Lowenthal, Bowden, Tuan, and other pioneers in the study of environmental perception.

Geography and Arena: Meaning, Metaphor, Milieu

The reasons for geography's contextualist turn during the twentieth century are manifold. Exploration and mapping of foreign lands were virtually complete; the task of further inventory and documentation was by now well anchored in the civil service or other institutionalized expertise of nations and regions. As in other lands, twentieth-century geographers developed styles of practice that differed from those of their mentors, who had sought grand generalizations and comparisons among vast regions of the world. They tended to limit their horizons of research curiosity to more circumscribed topics, usually those relevant to the home nation. They also felt the need to align their practices with the concurrent orthodoxies of scientific method and publicly acclaimed styles of scholarship. Vicissitudes of war and peace, economic growth and depression, and transformations of world demographic and political patterns had, by and large, rendered many classical texts obsolete or ideologically unpalatable. And, perhaps most significantly, the hegemony of a few imperial schools was certainly over by midcentury.

In retrospect, contextual awareness in geographic thought and practice echoed profound transformations in twentieth-century approaches to knowledge generally: the movement away from "spectator" to "participant" stances (Rorty 1979; Toulmin 1983). Evidence of philosophical reflectiveness among American geographers is rare, but the ideas of a few key writers have been germinal (Barrows 1923; Sauer 1925; Hartshorne 1939; Wright [1942] 1966). The mainstream pressed on with empirical field research and occasionally argued over methodological preferences for form or function, pattern or process (James and Jones 1954; James and Martin 1979). Tensions grew between those who emphasized the areal differentiation of the earth's surface (mosaic) and those who emphasized functional dynamics of spatial systems (mechanism), tensions no doubt heightened by wartime experiences and the Faustian prospects of postwar development and planning (PG, 186–95; Gould 1985).

Prior to the 1960s, most geographers would have positioned themselves in an "observer" stance on reality, aiming to achieve objectivity in their representations and explanations of phenomena. Issues of knowledge (*logos*), if discussed at all, were dealt with epistemologically. By the early sixties, however, there was a growing awareness of how human perceptions of reality had been filtered by different cultural groups, by different research instruments, and by different practical agendas (Wright 1966; Lynch 1962; Lowenthal 1961). The "social construction" of knowledge became a popular line of inquiry; queries arose about what or whose interests were actually being addressed in applied geography (Kates 1969; Hägerstrand 1970; Zelinsky 1970; Buttimer 1974; Mair 1986). This led to debates about power, language, and conflicts of interest between "insiders" and "outsiders" in particular situations (Antipode 1985; Racine and Raffestin 1983). Some sought to identify positivism as archvillain in a plot that had led the discipline squarely into the clutches of managerial interests (Samuels 1971; Seamon 1980). There were conceptually grounded arguments too. Following upon a few centuries of *ceteris paribus* orthodoxy, no doubt the time was ripe for those strong pleas from existentialist, phenomenological, pragmatist, and other sources that *ceteris,* for all practical purposes, were never *paribus* (Schütz 1944, 1973; Hägerstrand 1970; Buttimer 1974). As in other fields, *epistemological* questions yielded place or became transposed to *sociological* ones; the "foundational" concerns of the earlier (observation) phase yielded to "dialectical" ones (Stoddart 1981; Ley 1983).

By the late 1970s, a third phase was beginning. Geographers, like other social scientists, became aware of the fact that all were participants as well as observers, insiders or outsiders in different settings. Awareness also grew about the myriad ways in which conventional thought and practice had been filtered through the complexities of Western social experience. Reality became interpreted as an arena of events, of mirrors and masks, of texts reflecting contexts—a theater in which the antinomies of subjective and objective, normative and descriptive, internalist and externalist interpretations of science, were deemed anachronistic (Gregory 1981; Sugiura 1983). Some expressed the need for languages and symbols to facilitate a more open dialogue among different civilizations; others polished ever more sophisticated arguments about the futility of such a dream.

This wave of hermeneutic concern continued, but the late 1980s also witnessed a return to observation and a heightened humanitarian concern about ethics and intervention in the course of public life. The enormous scale and urgency of environmental issues and global change, in addition to dramatic realignments of demographic and political patterns, challenged the geographer to more field-oriented observation. Rapid advances in analytical technology, computer-based information systems, and satellite images changed the research horizon for the potential Leonardo da Vinci. As the philosophical strains of realism and pragmatism again recaptured audiences, disciplinary orthodoxy became less important than theoretical salience (Rowntree 1988; Ley 1989) and eventually problem resolution (Kates 1987).

Throughout these various transformations, the most consistent stream of contextualist endeavor in American geography was evident in its historical and cultural branches. Idealistic approaches to the study of landscape such as those inspired by Collingwood and others attempted to reconstruct the geographies of past times, seeing the landscape through the eyes of those who lived in them (Mead 1981; Guelke 1985). J. B. Jackson extended this to contemporary landscapes as well: "Every landscape is a reflection of the society which first brought it into being and continues to inhabit it" (1952, 5). Why not, then, "read" every landscape as text to be decoded in terms of the values of its human occupants? "To understand the landscape in living terms requires primary attention to the vernacular, to the environments of the workaday world. The motel, the franchised fast-food shop, and the contemporary American house seeking to accommodate new mobile and recreational lifestyles are as authentic examples of what vernacular means as the dwelling of a Pueblo Indian or Greek peasant" (Jackson 1976, 19).

During the 1970s and 1980s, this genre was revitalized (Salter 1978; Meinig 1979; Rose 1981; Sugiura 1983; Lewis 1986; Rowntree 1986). Landscape seen as the "sedimentation" of diverse forms of discourse allowed room for interpretations of various kinds (Norberg-Schulz 1980; Olwig 1984; Cosgrove 1984; Daniels 1988; Berdoulay 1988). It invited inquiry into culturally varying modes of symbolic transformation such as those inspired by Vico or Herder (Dainville 1964; Mills 1982), as well as semiotic inquiry into landscape texts and signs as products in their own right (Choay 1981; Marchand 1982).

It has surely been in questions of language and power, of semiotics and symbolism, that latter twentieth-century geographers have found the most challenging moment of interaction with structuralists (Gale and Olsson 1979; Gregory 1981; Dematteis 1985; Rose 1987). The deconstruction of inherited meanings, as suggested in the work of Derrida, deliberately sought to reveal the multiple and conflicting readings that could be made of particular texts (Derrida 1972). In the 1980s, Gunnar Olsson sermonized eloquently about the modern malaise in human geography: "Thing yields to process, stability to change, certainty to ambiguity, noun to verb, being to becoming" (Olsson 1984, 73). In sharp contrast with the previous generation's deeply held convictions about epistemological certainty and its image of the world as complex mechanism, this approach unmasked the extent to which *la condition humaine* "is one of the predicaments lived behind prison walls" (84).

It may well be premature to attempt an evaluation of this contextualist wave in geography. Its impact has been felt not only in terms of cognitive styles and choice of substantive foci of research but also in modes of vocational meaning. One of its major results has been a reaffirmation of *poesis* and the critically reflective element, long overlooked or trivialized in a profession that unequivocally favored scientifically based approaches to *logos* and *ergon*. Ontological issues are again discussable; modes of discourse themselves have become matters for critical reflection (Folch-Serra 1989; Karjalainen and Vartainen 1990). Substantively, there has been a renewal of research curiosity about relationships between humanity and environment, the meanings of place and landscape for human creativity and health (Hudson-Rodd 1991). Flowing from this, environments themselves reassume significance as contexts for life, cultural identity, and historical legacy—a long-overdue rediscovery after a few generations that focused on space as a tabula rasa on which diverse models of economic or technological rationality could be tested. Within the span of a few decades, even mainstream approaches to *logos* and *ergon* have changed. Those blithely naive confusions of descriptive and normative discourse which characterized some postwar ventures in applied geography are now less palatable to audience and sponsor alike. The harvest may not yet be ready, but at least there are certain fruits of new (or rediscovered) awareness evident among practitioners of human geography today, and these bear profound implications for future research on humanity and its terrestrial home.

> *Geosophy:* awareness of cultural relativism in the ways in which human groups construe nature, resources, society, space, and time. Curiosities about differences in the "geographical sense" of various peoples, which created a considerable body of literature on environmental perception and behavior.
>
> *Temporality:* awareness of the inextricable bonds of time and space in all phases of geographical curiosity. There has been a renewal of interest in different temporal scales involved in everyday experiences of environment which is distinct from the historical geographer's traditional insistence on the "contingency" of human events (Hägerstrand 1971; Mårtensson 1979).
>
> *Relativity:* awareness of the inevitable ripple between observer and observed (Heisenberg/Bohr) has heightened consciousness of how research instruments themselves—conceptual as well as analytical—selectively focus inquiry.
>
> *Agency versus Structure:* awareness of the social dimensions of academic lifeways has led to explorations in the history and sociology of disciplinary practice, to discussions about paradigms, to definitions of science, normal and extraordinary, and perhaps most significantly, to awareness of mirrors and masks in the conduct of everyday life in the academy as well as in the marketplace.
>
> *Hermeneutics:* awareness of cultural biases in the design of research as well as in the conclusions derived from them has led to debates over

ideology, knowledge, and power, or, for some, to acknowledgments of the hermeneutic circle. The researcher has begun to acknowledge his/her role as participant in, rather than observer of, reality.

Stimuli for these new kinds of awareness have come from interactions between geographers and scientists, philosophers, and historians of ideas. Thus far, contextual approaches have appealed more to practitioners of human rather than physical geography. Global concerns of the nineties have again afforded opportunity for reintegrating human and physical branches of inquiry. There is a growing awareness of interdependencies among nations and regions, not only in economic and technological terms, but also in the bioecological consequences of inherited lifeways (Buttimer et al. 1991; Zonneveld 1986). With this has come an extension of intellectual horizons in time as well as in space, concern about the past and also about the future. For those who have endeavored to incorporate these new kinds of awareness in their practice, there still remain certain challenges. Of these, three currently call for creative energies; first, how to interpret the Protagorean motto "The proper study of mankind is man"; second, how to relate narratives on "events in context" or on "landscapes as texts" to the wider issues facing humanity and world; third, how to determine what potential message the Western world has to offer on the present-day global crisis in the interactions between anthroposphere, biosphere, and geosphere.

The challenge of linking insight into the experiences of individual persons or places with that of the more general experiences of regions and cultures remains fundamental (Daniels 1985, 1988; Rowntree 1988; Sterritt 1991). Some still regard the aims of *Wissen* and *Verstehen* to be forever irreconcilable. Others claim that knowledge is always socially constructed, and therefore modes of disciplinary practice should be understood in ethnographic rather than epistemological terms. Impasses over the One and the Many, the ideographic and the nomothetic, linger strongly within human geography (Ley 1989).

Among the advocates of contextual approaches, too, a key distinction can be made between those emphasizing *events* and defining contexts in terms relevant to those events and those giving primary attention to *contexts* (physical, ecological, functional) and searching for a potentially universal grid or explanatory theory to account for all possible events. In a sense, the old distinction between the spirits of Ptolemy and Herodotus has taken on a new significance. On one side stand the would-be guardians of local cultures and the integrity of places who disdain universal laws of context. On the other side are the would-be managers of universal systems to guide and plan human behavior and activities on our fragile and threatened planet.

The contextual approach has yielded its best results when the focus of inquiry has rested on particular events or periods. It permits appreciation of the complex intertwining of ideas and practices with general societal and environmental circumstances in which they are disciplined and implemented. Gaining much better insight into particular "nows," one simultaneously risks losing the threads of historical flow or possibilities for cross-cultural comparisons and contrasts. While the humanist spirit would rejoice at this growing contextual sensitivity, it would also wish to affirm the values of cross-cultural understanding and historical depth (Harris 1978; Daniels 1985; Folch-Serra 1989). Finally, like their colleagues in social science and humanities, geographers have also tended to define "context" in anthropocentric terms, focusing on humanly constructed frames of reference of laws, structures, and artifacts surrounding events. Vico's appeal for a New Science has been heard. The most demanding challenge for the geographer today is how to incorporate nature and the biophysical environment in descriptions of context while avoiding the resuscitation of environmental determinism.

Narcissus, Twilight, and Arena

The cultural diversity of the anthroposphere has posed interminable puzzles for scholars and missionaries, emperors and traders, through the centuries. The written record is replete with examples in which scholarly opinions have apparently been harmonized with the ongoing *Zeitgeisten* of sponsors and audiences. At times environmental determinism has held sway, at times hierarchical classifications; and more often than not, lifestyles and cultures have been evaluated in terms of their deviations from one's own cultural norm. To raise questions about taken-for-granted world images has always been a hazardous affair; one might dare to do so from the periphery, from one's exploring vessel, from exile, or from a country villa during retirement years.[5] And then one's aim might vary from that of evoking a nostalgia for some ideal model of society from which one's people has deviated, to that of imagining alternatives, for example, along the lines of "simpler" (or less technologically sophisticated) people discovered in some other part of the world.

In historical perspective, most interest is aroused by those instances where Narcissus creates not only a sense of cultural relativism (that people in different environments construe their world in different ways) but also a sense of critical self-understanding. Many of the great satirists in history have been deemed successful (albeit sometimes retroactively) with the former, but only a few have succeeded with the latter. Through the centuries, great empires have recorded ethnographic accounts of the cultural diversity of humankind in the categorial frameworks of a world mosaic, mechanism, or organism; as long as they could be articulated in the language of the audience or sponsor, and not therefore stretch imaginations too far beyond the established routines of one's home world, such accounts made for pleasant conversation and edifying pedagogy. It is at the point where awareness of cultural relativism could be combined with critical self-understanding that fundamental changes in the taken-for-granted ways of thought were conceivable. The world as arena eventually implies no hierarchical ranking of cultures according to their resources, skills, corporeal beauty, crafts, musical styles, or demography: ultimately, in the arena, all potentially have an equal chance of playing out their *genres de vie*.

Awareness of context, and of the manifold bonds between language, culture, and scientific ideas, was certainly not an American or even a twentieth-century invention. Nor is context necessarily ignored in the other world-views sketched here. In Western intellectual history, curiosity about events-in-context has characteristically arisen in the twilight zones of civilizations, at times and places where long-established certainties of thought and life had begun to crumble, and when some, at least, glimpsed possible alternatives. Such reflection has also characteristically emanated from the periphery rather than the center of regimes, and among explorers of marginal or remote regions.

> The edges of any landscape—horizons, the lip of a valley, the bend of a river around a canyon wall--quicken an observer's expectations. That attraction to borders, to the earth's twilit places, is part of the shape of human curiosity. And the edges that cause excitement are like these where I now walk, sensing the birds toying with gravity; or like those in quantum mechanics, where what is critical straddles a border between being a wave and being a particle, between being what it is and becoming something else, occupying an edge of time that defeats our geometries. (Lopez 1986, 110)

The world as arena might be regarded as that world most congenial for Narcissus-as-pilgrim, and one that also offered scope for Phoenix to arise from the ashes of oblivion. The contextualist has at times been moved by malaise over contradictions between conventionally held beliefs and the actual facts of life and landscapes. Often, too, there has been nostalgia for modes of thought and being more closely attuned to direct (sensory and emotional, as well as practical) contact with the natural world and to one's fellow humans. Trickling along the edges of mainstream Western journeys toward pure reason and theoretical certainty, there are streams of nostalgia for the pre-Socratic grasp of fusion and integrity in the reciprocity of mind and matter (*nous/physis*), dynamism and mutational flux, an immanence within physical matter as well as within human consciousness, that creative power and spirit which in post-Socratic times would be abrogated by a transcendent deity or by scientific truths. Bruno's poetics about an infinite universe confronted the fixed categories of Scholastic ontology; Montaigne's autobiographically based essays raised questions about ecclesiastical and cultural myopia; Thoreau's *Walden,* like Humboldt's *Cosmos,* appealed to nature as source of alternative foundations for human knowledge and ways of life.[6]

"In the reprieve at the end of a day," Lopez wrote, "in the stillness of a summer evening, the world sheds its categories, the insistence on its future, and is suspended solely in the lilt of its desire" (1986, 66). Here is a worldview that should be least amenable to becoming part of a Faustian *hohen Streben,* but one that might deny Mephistopheles a victory in the final act of the Faustian drama. The contextual turn has begotten anarchist élan, as well as instrumentalist rule of thumb; its twentieth-century flowerings have been in the understanding of cultural relativism, relativity in science, and the social construction of reality. It has also unmasked the fallibility of traditional doctrines about the universe and its human occupants and has revivified the Heraclitean concept of never stepping into the same river twice.

Pleas for contextual sensitivity in Western humanity's modes of thought and life have heralded Phoenix moments of discovery, higher-order integrations of knowledge and understanding about humanity and earth. Most remarkable among these have been advances in physics, astronomy, and the sciences of the biophysical world from Nicolas of Cusa to Niels Bohr. No less emancipatory have been the affirmations by poets and anthropologists of the uniqueness of people and places and the importance of aesthetics, faith, pleasure and pain, passion and reason, in human journeys, from Montaigne and Vico to Geertz and Lopez. The solitary searches of artists and mathematicians, too, from Picasso and Van Gogh to Gödel, Escher, and Bach, have preceded moments of discovery in language and human communication (Hofstadter 1979). The twilight and the circadian phenomenon of rhythmic alternation between darkness and light, repose and action, offer the best context for this metaphor of world as arena.

As the millennium of Pax Romana faded around the tenth and eleventh centuries, and diverse peoples from Eurasia and North Africa suddenly invaded Mediterranean lands, a great questioning fermented. During the latter part of these so-called Dark Ages, scholars and sages from Cordoba to Clonmacnoise traversed Europe with unconventional ideas. Elements of Arab and Celtic philosophy, medicine, and geography, trace elements at first, ultimately transformed the sedimented certainties of Roman (and Platonic) Christendom. Without such ferment, such contextual curiosity about events and places, the Renaissance Phoenix might never have been possible. Fresh perspectives on the anthroposphere were then proclaimed, affirmations about the integrity of the individual human person, that potential microcosm of the universe, with both right and responsibility to fulfill his or her highest potential. Later, in the eighteenth-century Enlightenment optimism about a mechanically steered universe, contextualist questioning yielded a rediscovery of nature, a reaffirmation of the biosphere, and the wisdom recuperable from careful scrutiny of how natural events actually unfolded. The twentieth-century contextual turn, following virtually a century of atrocities in imperial expansionism, environmental destruction, and totalitarianism, elicited new forms of sociality among humans and other living creatures and underscored especially the key role of the socio-technosphere in humanity's modes of making its terrestrial home. At the close of the twentieth century, one finds a return to the Hellenic dilemmas about the noosphere itself, the ambiguities of language and the persistent interplay of rationality and irrationality in the world of thought and its relationships to *physis,* humanity's terrestrial home.

A poignant problem faces the contextually sensitive geographer today. Arena, like organism, is a root metaphor that has served best for *poesis* and *paideia.* Beyond those antinomies of ivory tower versus applied research, the social organization of research and teaching reflects its national sponsorship. Contextually grounded analyses of problems, for example, those of environmental hazards or catastrophes, are welcomed as long as results are easily translatable into terms understandable in the context of policy and action. A politically understandable *logos* is invited as grist for the mill of democratically appointed political *ergon.* While *arena* is promising as a guiding metaphor in the search for a holistic description of events, its recipe for translation between descriptive and normative statements demands a relativistic attitude toward truth and being. Each particular problem invites its own particular solution, an operational theory of truth offering *logos* for that *ergon,* but no claims are made for this as a paradigm or model solution applicable thereafter to all other conceivable problems of that genre within that nation's zone of relevance. This ad hoc operational approach to truth, the leitmotif so essential for the root metaphor of world as arena, is scarcely reconcilable, however, with the overt aims of politically structured expertise for policy and planning in post-frontier settings.

If Voltaire's *Candide* (1759) epitomized the challenge for the rising star of modernism, then Borges' *The Book of Sand* (1977) epitomized the dirge for its wake in the latter years of the twentieth century:

I opened the book at random. The script was strange to me. The pages, which were worn and typographically poor, were laid out in double columns, as in a Bible. The text was closely printed, and it was ordered in versicles. In the upper corners of the pages were Arabic numbers. I noticed that one lefthand page bore the number (let us say) 40,514 and the facing right-hand page 999. I turned the leaf; it was numbered with eight digits. It also bore a small illustration, like the kind used in dictionaries—an anchor drawn with pen and ink, as if by a schoolboy's clumsy hand.

It was at this point that the stranger said, "Look at the illustration closely. You'll never see it again."

I noted my place and closed the book. At once, I reopened it. Page by page, in vain, I looked for the illustration of the anchor. "It seems to be a version of Scriptures in some Indian language, is it not?" I said to hide my dismay.

"No," he replied. Then, as if confiding a secret, he lowered his voice. "I acquired the book in a town out on the plain in exchange for a handful of rupees and a Bible. Its owner did not know how to read. I suspect that he saw the Book of Books as a talisman. He was of the lowest caste; nobody but other untouchables could tread his shadow without contamination. He told me his book was called the Book of Sand, because neither the book nor the sand had any beginning or end."

The stranger asked me to find the first page.

I laid my left hand on the cover and, trying to put my thumb on the flyleaf, I opened the book. It was useless. Every time I tried, a number of pages came between the cover and my thumb. It was as if they kept growing from the book.

"Now find the last page."

Again I failed. In a voice that was not mine, I barely managed to stammer, "This can't be."

Still speaking in a low voice, the stranger said, "It can't be, but it is. The number of pages in this book is no more or less than infinite. None is the first page, none the last. I don't know why they're numbered in this arbitrary way. Perhaps to suggest that the terms of an infinite series admit any number."

Then, as if he were thinking aloud, he said, "If space is infinite, we may be at any point in space. If time is infinite, we may be at any point in time." (1977, 118-19)

This is a far cry this from the daring deliberations of a Peirce, a James, a Dewey, or even a Pepper. American pragmatism's operational or instrumental theory of truth was still anchored in implicit beliefs that there were some foundations for knowledge. Pepper's contextualism thus offers only partial guidance toward an exploration of the world as arena. For through the windows flung open to admit the root metaphor of the event-in-context has blown a gale-force storm, which has invaded virtually all domains of

Western thought and experience. Some have set about rearranging the furniture or redesigning it according to the models of mosaic, mechanism, or even organism; others have regarded the house itself as something blown down by the wind and conceive of little alternative than to allow everything to fly away, there no longer being any commonly held criteria for distinguishing between reality and illusion in thought or in life. If the sagas of the other three world-views, mosaic, mechanism, and organism, could be recounted in terms of their journeys from classical to modern times, then that of arena points inevitably to the dramatic dilemmas of postmodernism.

Discordant melodies now playing in the world arena thus echo perhaps a speeding up of those cyclical tensions among Phoenix, Faust, and Narcissus. Now that contextual sensitivity has become more or less part of the sine qua non of scientific protocol, several dilemmas must be confronted. Dramatic events, and equally dramatic transformations of context since the 1940s, have clouded the prospects of explanation which were so optimistically held at midcentury. Assumptions about universal common sense and laws of nature, for instance, have themselves been controversial. An all-pervasive nihilism has shattered those horizons of hope which characterized pragmatist writings prior to World War I and which persisted, to some degree, throughout the interwar years. These latter years of the twentieth century have witnessed a return to utilitarian and operational guidelines for life and thought; the social climate is one where orthodoxy and dogmatism capture increasing acclaim. Somewhat paradoxically, in academic circles, the world of truth has become for many a Borgesian book of sand, an arena of relativity and relativism, of Heraclitean flux and black holes, a world in which virtually all inherited notions about reality are inextricably entwined in a hermeneutic circle.

Concluding Hopes: Narcissus Awake

THE FOREGOING CHAPTERS record an extended visit to the muses of Helicon. At the outset there was the hope that Narcissus, once critically aware of the tensions between Phoenix and Faust in the saga of his own self-becoming, could emerge from the pilgrimage with a fresh kind of self-understanding and a readiness to explore broader horizons of life and thought. The pilgrimage began with queries about the integration of knowledge, and especially about a rethreading of the strands of physical and human geography. Along the way, reflections on various career journeys have sought common denominators on which mutual understanding might be possible among specialists in diverse fields of science and the humanities. This concluding chapter now summarizes ways in which these interpretive themes elucidate the question of integration, and their implications for the practice of geography.

In historical perspective, the term *integration* might best be understood as a verb rather than as a noun, as a process rather than as a state, as a tendency in various spheres of life and thought which is dialectically opposed to individuation or dispersion. The themes of meaning, metaphor, and milieu allow for an initial unraveling of this process from a contextual vantage point. Instead of focusing on individual subjects (authors) and their dispositions, the framework highlights practices, thought styles, and substantive research interests. It enables one to consider integration in geography in terms of three main spheres or levels of concern: disciplinary organization and professional practice (structural level), intellectual traditions (conceptual level), and relevance (societal level). In the light of virtually all the career stories examined in this study, the central issues identified throughout all three levels are those of *integrity* versus *integration*.

Structural questions, for example, on the status and interactions among subdisciplines and the relative priorities placed on arious practices, can be elucidated via categories of meaning: *poesis, paideia, logos,* and *ergon*. It is, after all, on this level that individual geographers, research teams, and departments make their day-to-day contributions to the discipline and negotiate for sponsorship and audience. Administrative and managerial interests might find it convenient to establish priorities among these practices, integrating geographical expertise within the overall division of labor dictated by the political expediency of the times. The integrity of disciplinary practice, however, demands a flexibility to changing educational needs and to new substantive research challenges. Ideally each individual, department, or research team, given its resources, aims, and context, should assume responsibility for designing and adapting its agenda to such changing demands.

On conceptual questions, the integrative process raises issues of cognitive style: whether there might be a body of theory or arsenal of methods which could provide an intellectual core for all branches of geographic inquiry. On this level, the root metaphor approach enables one to evaluate the relative strengths and limitations of integrated versus dispersed images of reality. Clearly, the integrated approaches of *organism* and *mechanism* have invited research of wider scale and have apparently enhanced the status of geography within the academy. The dispersed approaches of *mosaic* and *arena* have yielded more sensitive accounts of life and landscape at local and regional scales of inquiry. The integrity of any root metaphor's truth claims has been threatened when its products are submitted to criticism in terms of another. The integrity of geographical thought as a whole has been compromised whenever any one of these is flaunted as the sole basis for orthodoxy of cognitive style in the discipline.

At the level of societal relevance, integration points toward geography's actual or potential engagement in the public (environmental or social) issues of its time and setting. The categories of milieu concern, *identity, order, niche,* and *horizon* offer useful foci of attention. There are clearly major differences between "insider" and "outsider" perspectives on each of these interests; more significantly, the integrity of places, regions, issues, and resources has often been compromised when they have become integrated within national or regional schemas of analysis or policy control.

Early on in this project it seemed indeed that the question of integration could be regarded as the managerial challenge, while the scholar's concern might be primarily that of integrity (Buttimer 1985b). The process of scientific inquiry could be seen as always seeking higher (namely, better integrated) levels of understanding, while simultaneously seeking to address societally relevant issues. To rest with any particular model of integrated knowledge, or to allow one's research curiosity to be fully integrated with any particular societal context, could spell sclerosis. The current implication is that geography as a scholarly discipline can only remain creative and seek integrity when there is ample scope for a playing out of those tensions between integrative and dispersive forces, a dialectic of stability and innovation, of security and adventure, on all three levels of its practice. For once the integrative process reaches fulfillment—either in terms of institutional structures, paradigmatic certainty, or public relevance—then scholarly energies are harnessed toward routine operational tasks. Integration as *fait accompli,* in any or all of those senses suggested in my framework, might be tantamount to the kiss of death for intellectual creativity.[1]

This conceptual distinction between integration as a managerial challenge and integrity as the scholarly one flows logically from the interpretation of career accounts described in Chapter 1, where tensions were noted between the personal dispositions and talents of individuals and the priorities set by the structures in which they worked. While this distinction remains useful at a general level, a closer look at contemporary university settings reveals a much more complex picture. Unlike previous times, there is now, within and without the university, a veritable Frankenstein's monster of institutional arrangements whose sole *raison d'être* is to negotiate the terms of scholarly practice. The time and energy expended on protocol and procedure, however well intentioned, inevitably detracts from the essential work for which one chose the intellectual life in the first place. Functional specialization within sectors of managerial expertise dismembers rather than unifies; far from facilitating integration, it actually leads to a disintegration of scholarly life and thought.

External influences, screened through these mediating layers, amplify their effects on the perceived value and meaning of disciplinary practice. The growing dominance of *ergon* and persistently restrictive approaches to *logos* have undermined and trivialized *paideia* and *poesis*. How priorities change is a complex story, but during the latter half of the twentieth century they have been most effectively steered by the political economy of research grantsmanship. Consider the time and energy devoted by scholars to writing and screening research grant proposals. Few would question the fundamental values of promoting science and improving the quality of research: otherwise, why would they have chosen academic careers? But the searing tensions between ethos and structures today can no longer be described in terms of "insiders" and "outsiders," and of scholarly versus managerial interests. Rather, we find ourselves entangled and often lost in the labyrinthine maze created by ourselves, even with the noblest of intentions.

This progressive integration of academic life into the public sectors of national economies constitutes one of the roots of the humanist *cri de coeur* for freedom of thought today. While world geopolitical realities show how erodable national frontiers are, the practice of research has become more solidly entrenched within, and dependent upon, national budgets. This poses special problems for geography, whose horizons of research curiosity should ideally be global in scope. Together with colleagues in history and the humanities, the contemporary geographer needs equally broad horizons of time. Research policy, which favors short-term, present-day-oriented, and strictly circumscribed projects, militates against the eventual production of scholarly works that could lead to the kinds of discovery and understanding which are needed in facing the challenge of global existence tomorrow. Market forces, too, exerting influence on the diffusion of scholarly products, reinforce cultural and territorial biases that affect the availability of fresh ideas. Late twentieth-century trends suggest that the scholarly content or the provocative merits of a work weigh far less than how well it is expected to sell.

Some historical perspective is therefore essential if one is to escape the claustrophobia ensuing from an exclusively structuralist or synchronic interpretation of geographic thought and practice. The second trilogy of themes—Phoenix, Faust, Narcissus—facilitates a diachronic view of the changing contexts of disciplinary practice. The practice of geography has been profoundly influenced by changes in national mood and fortune. Geographers had little difficulty maintaining an integrated field as long as it was fueled by national self-confidence or altruism. At moments of waning national self-images, queries were raised about the discipline's identity and images of the world. Beyond national frontiers one can discern the shadow of Prometheus. That same Faustian spirit which at one time fostered imperial expansion (political, commercial, and intellectual), and later sustained visions of economic development and rational planning, today continues to build—for humanity's sake—mazes of institutional forms which swell the ranks of officialdom in what is euphemistically called the public sector. Narcissus in these postmodern times surely longs for Phoenix.

Each fresh discovery in humanity's geographical understanding has heralded changes in its self-image, hopes, and fears. As suggested in Chapter 6, Phoenix moments in Western intellectual history have been those that emerged after Narcissist "twilight longings" for new levels of understanding humanity and its terrestrial home. Classical Greece explored the nature of creation as a whole, observing interconnections between mind, society, nature, and the gods, and proclaiming *reason* as the distinctive quality of humanness. The convergence of Hellenic, Judeo-Christian, and Arab imaginations in medieval Europe led to an image of the world as mosaic of forms. The Renaissance raised the issue of human dignity, celebrating the human person as a microcosm of the universe, and rational scientific inquiry as capable of explaining it; thereby it loosened the bonds of ecclesiastical dogmatism and widened the geographical horizons of medieval Christendom. Its Faustian energies reaped a harvest in the science and technology of Enlightenment times, where imaginations were steered toward an image of the world as machine. Romanticism, rebelling against the scientific, totalizing rationality of the Enlightenment and heralding a view of the world as organic whole, rediscovered nature as primeval and sacred force. The twentieth century contextual turn, echoing many previous turnings to the here-and-now, revealed the intricate interweaving of thought and life, particularly the role of society in structuring relationships between humanity and world. Thought itself would come to be regarded as a social product. Late twentieth-century scholars could no longer consider themselves as mere observers of the world; rather, they were to regard themselves as participants in an arena of spontaneous events.

In each of these periods there was a clearly identifiable emancipatory élan, often aiming at deliverance from confining orthodoxies, structures of thought, politics, or material conditions, at times uttering a prophetic note about future possibilities.[2] Each, too, witnessed a will to build structures and institutions, bequeathing its own legacy of Faustian forms and unresolved tensions to its offspring. In the subsequent attempt to understand and transcend those structures, during those dark moments before the dawn of a new Phoenix, there was the ardent longing of a Narcissus.

Catalysts for critical reflection abound during these latter years of the twentieth century: frustrations over those Faustian fences that today surround thought and life, disillusionment over the ironic contradictions of word and action in many idealistically motivated social movements, and maybe a lingering fear about those Spenglerian prophecies of the decline of the West. Given such an inheritance, one should naturally expect expressions of vulgar Narcissism, well-justified outpourings of satire, and brilliantly articulated autopsies of "realized ideals." But the very experience itself can evoke a thirst for the other face of Narcissus, the pilgrim gazing toward new horizons that could reveal a Phoenix.

This retrospective glance at the Western story reveals undeniable evidence of progress in the human understanding of earth and world, of resilience despite disaster and prostitution to the structural imperatives of various eras, and of the malaise felt, especially in the twentieth century, over all that impedes the free circulation of thought and life. Having passed through a reflective and Narcissist phase, the humanist spirit again desires deeper comprehension, from the heart of the atom to the mysteries of the universe. Still, the dramatic contrast between humanist and scientific modes of knowing often remains unchanged. Individual scholars within both worlds express concern about the wider picture, but from different perspectives: scientists still wish for integration of knowledge; humanists, wholeness of being.

These two distinct images of integration coexist within contemporary geography. They are rooted in two very different modes of understanding: the rational-scientific and the mythopoeic. The mutual hostility (or indifference) between these two conceptions of knowledge is perhaps a deeper problem than the institutional separation of physical and human branches of the field. Being housed under one roof, even as they still are in many departments today, does not necessarily mean there is much communication or even semblance of integrated practice. The overwhelming consensus among would-be guardians of research quality is that the rational-scientific mode should hold hegemony in disciplinary inquiry. With this has come a peculiar paradox: a penchant for describing the earth in literal, materialist, and reductionist terms, on the one hand, yet a penchant for a totalizing, generalizing approach to normative action on the other. For all its conquests in scientific explanation and all the technological spin-offs, such a practical style has blinkered our vision of the full range of environmental experiences of other civilizations.

For the human geographer, especially in faculties of social science, exploration of the earth's surface has implied inventories of resources, population, and habitat, as well as elegant models of spatial organization and circulation systems, namely, descriptions of the earth as potentially exploitable and controllable *Lebensraum*. Physical geographers, especially those affiliated with faculties of natural science, have interpreted those same landscapes, *ceteris paribus,* in terms of tectonic, erosional, hydrographic, or bioclimatological systems. This has meant that neither "human" nor "physical" geographers felt equipped to deal with so-called environmental issues when these eventually surfaced as matters of public concern. Still relatively ignored, poorly glimpsed, or possibly misunderstood are everyday lived realities—the spiritual, emotional, moral, and symbolic meanings of nature for people in other civilizations and other milieus.

The closing years of the second millennium A.D. may indeed be time for Occidental civilization to reflect on its twin foundations in Hellenic philosophy and Judeo-Christian theology, remembering its own forgotten strains of caring for Gaia, and at the same time being open to understanding mythopoeic and experientially grounded accounts of other civilizations. Reflections on myth and symbolism in cross-cultural perspective afford such an opening. At the antipode of Greece and Rome, Australian Aboriginals tell the following story of how the world came to be:

> In the beginning the world lay quiet, in utter darkness. There was no vegetation, no living or moving thing on the bare bones of the mountains. No wind blew across the peaks. There was no sound to break the silence. The world was not dead. It was asleep, waiting for the soft touch

of life and light. Undead things lay asleep in icy caverns in the mountains. Somewhere in the immensity of space Yhi stirred in her sleep, waiting for the whisper of Baiame, the Great Spirit, to come to her.

Then the whisper came, the whisper that woke the world. Sleep fell away from the goddess like a garment falling to her feet. Her eyes opened and the darkness was dispelled by their shining. There were coruscations of light in her body. The endless night fled. The Nullarbor Plain was bathed in a radiance that revealed its sterile wastes.

Yhi floated down to earth and began a pilgrimage that took her far to the west, to the east, to north, and south. Where her feet rested on the ground, there the earth leaped in ecstasy. Grass, shrubs, trees, and flowers sprang from it, lifting themselves towards the radiant source of light. Yhi's tracks crossed and re-crossed until the whole earth was clothed with vegetation.

Her first joyous task completed, Yhi, the sun goddess, rested on the Nullarbor Plain, looked around her, and knew that the Great Spirit was pleased with the result of her labour.

"The work of creation is well begun," Baiame said, "but it has only begun. The world is full of beauty, but it needs dancing life to fulfil its destiny. Take your light into the caverns of earth and see what will happen." (Reed 1980, 11)

A more conventional scientific account might read as follows:

The struggle for survival in the final analysis embraces all means and methods to obtain energy or to escape being converted into it. If our reconstruction of the earliest living things is correct, we must conclude that they led a passive existence, floating aimlessly in the seas among a variety of carbon-rich but lifeless molecules. By means of purely random movements, the living organisms occasionally contacted the nonliving molecules. The two entities became attached by chemical bonds and there was a consequent increase in size and duplication of structure. This purely passive method of keeping alive was adequate for a time and continued as long as there were sufficient stores of suitable carbon compounds in the environment. Only when the original energy sources were depleted or when local living space became crowded did competition begin. At this stage Darwinian evolution took over.

If it were not for the process of photosynthesis, all life might have disappeared from the earth at an early stage of evolution. Photosynthesis was the only method of obtaining life-supporting energy that was capable of operating on a sustained basis because it utilized the one inexhaustible energy source—solar radiation.

In green (chlorophyll-bearing) plants, photosynthesis takes place when sunlight is utilized to convert carbon dioxide and water into energy-rich carbon compounds, with the release of oxygen as a byproduct. The simplified reaction in chemical terms is as follows:

$$6CO_2 + 12H_2O \text{ —solar radiation— } C_6H_{12}O_6 + 6O_2 + 6H_2O$$

> When properly combined . . . the carbon compounds and the oxygen yield the energy that supports animal metabolism. Organisms that were most effective in combining oxygen and food were favored in the struggle to survive and were able to displace their less adaptable associates. (Stokes 1973, 222–23)

Like all interpretations of land and life, both of these accounts nourish the geography of places, but neither can supply all the necessary ingredients for a completely integrated account. Other accounts are indeed conceivable, but would the collection add up to an integrated picture? From what inquiry standpoint is it possible to relate these interpretations?

Western history suggests that impulses to transcend one's taken-for-granted modes of thought and practice nearly always derive from extra-scientific sources, such as building a nation, liberation from oppression, or visions of a better world. Gaia, déjà vu for many, could still hold poetic appeal as horizon for both human as well as physical geography.[3] Like Yhi in the aboriginal story, the idea of a living and resilient planet now dawns on the Nullarbor Academic Plain once glaciated by imperialistic science and technology. It certainly affords ample challenge for joint exploration into those perennial human interests of identity, order, niche, and horizon; and it could reaffirm the geographer's role in promoting better understanding among the peoples of the earth.

In retrospect, Phoenix moments for human thought—illuminative stars in the noosphere—have burst through the Faustian fences and frameworks within which managerial interests sought to contain them: fresh understanding of mankind's terrestrial home has led to fresh understanding of mankind itself. In defiance of the traditional limits to their disciplinary niches, geographers have dared to look inward, exploring cultural differences in environmental perceptions and experience, where they encounter colleagues in the humanities who dare to look outward, freely exploring ground traditionally trod only by natural science and theology (Matthiesson [1979] 1987; Lopez 1986; Eco 1986). The occasion for joint exploration into the mystery of human creativity and wisdom in humanity's modes of dwelling has come. New alliances between physical and biological sciences as well as between them and the human sciences are taking shape within the academic world (Lovelock 1979; Prigogine and Stengers 1984). One can scarcely again consider "matter" as dead or nature as a complex of blind forces; rather, one is discovering the complex and dynamic wisdom written into the nature of the universe (Lovelock 1979),[4] and the basic bonds between humans and fellow living creatures on the earth (Kohak 1984; Lopez 1986).

The late twentieth-century map of humanity shows that Euro-America and its legacy of Hellenic and Mediterranean models of *humanitas* is but one corner in an evolving noosphere. Yet analogies with pre-Renaissance Mediterranean worlds is surely striking (Eco 1986, 59-86). The contemporary world map of power has taken on radically new dimensions due to transformations in technology and trade. The explosion of information and the transcontinental circulation of people, commodities, and ideas raise new possibilities for thought and life, if we dare transcend our inherited orthodoxies, or at least critically reflect on tensions between their spirit and their letter. It was just such transformations that confronted the imaginations of Renaissance pioneers, many of whom were jack-of-all-trades poetic types (see fig. 13) and therefore eager to stay abreast of new developments in several realms of human becoming. What the retrospective glance deems as "revolutionary" could scarcely have been possible if these diverse aspects of humanness had not been regarded as mutually enriching, either in dialogue among human individuals or within the individual person.

Implications for the current practice of geography are obvious. While each of our practices demands its own level of specialization, so, too, each needs to be orchestrated with the others. Autopsy of inherited structures and externally imposed priorities may shed light on how our present fragmentation of effort has come to be, but to suggest that only structural changes are needed to clear the way for tomorrow's styles of practice would surely betray a lack of imagination. The humanist's emancipatory hope would be for the scholarly community to begin at home, critically evaluating inherited priorities and bravely experimenting with new ways to orchestrate intellectual energies.

Communication and mutual understanding, the very tissue of the noosphere, are among the greatest challenges facing human geographers today. And this effort involves emotion and will, as much as cognitive brilliance, technical ingenuity, or the design of media. Ultimately it redounds to an expansion of heart and spirit to embrace the twofold challenge facing humanity's dwelling in its terrestrial home in the late twentieth century: sociality and ecology. *Zoon politikon* today faces the challenge offered by the spatial juxtaposition of culturally diverse peoples: to let these often involuntary movements and convergences of humans become the springboard for new creativity in politics and social life, rather than problems to be solved by social engineering. Ecology also, in the original meaning of the term (from *Oikos,* meaning home), calls for a sense of how life as a whole functions. Today's chorus of protest against environmental destruction has indeed been heard, but the Faustian fashion in which problems are being addressed serves more frequently to fragment than to unify efforts toward ameliorative action. Geographers have much to offer through cross-cultural and historical evidence from mankind's experiences in sociality and ecology. Attuned to the emancipatory role, which humanist thought has played historically, too, they could reiterate at least the essential message that human reason cannot function without hope. Gaia's human envelope, the anthroposphere, needs to be understood as a drama more complex than simply a battleground of ecological versus economic rationality; rather, it needs to be understood as *oecoumene,* home for mankind, a species that urgently needs to rediscover the art of dwelling.

Much spleen has been vented on the Western world's anthropocentric perspectives, with critics usually implying alternatives such as the deification of nature, the social collective, or the reification of ideologies playing out their purposes with humanity and nature. Perhaps the West, far from overestimating *humanitas* has grossly underestimated it. The recovery of the human subject, the recognition of human agency as an integral part of the lived world, and the creativity again apparent in peripheral and previously marginalized regions all suggest fresh potential for human geography today (Folch-Serra 1989). One is challenged not only to regard humanity and earth in global terms but also to understand the ecological and social implications of a world humanity now planetized (Teilhard de Chardin 1955, 1959). The need for cross-cultural and comparative research implies an extension of time horizons on the history of the earth and human occupants to date, but also reflection on the potential future of Gaia (Needham 1965; Fraser 1975). Once aware of the strengths and limitations of our inherited conventions, Western scholars could be catalysts for a more mutually respectful exchange among peoples about the human experience of nature, space, and time. Mutual understanding among the earth's diverse civilizations has become, after all, not just the poetic dream of ivory tower scholars but an urgent social imperative in our day.

The recovery of humanism in geography could herald a potential Phoenix, emerging from the ashes of former tyrannies—methodological, epistemological, or ideological—in all facets of disciplinary practices. As a stance on life, humanism is one that welcomes the creative potential of individuals and groups to deal with the surface of the earth in responsible ways. Nor is human creativity confined to the intellectual sphere: it involves emotion, aesthetics, memory, faith, and will. As Phoenix, then, the humanist turn in geography should refuse to be contained, named, and claimed by Faustian structures. It can inspire practitioners of physical, economic, cultural, or social geography, and should perhaps not invest too much energy in staking claims for becoming a special branch of the field.

Humanism, this book suggests, is leaven in the dough and not a separate loaf in the smörgåsbord of geographic endeavor. The emancipatory élan recuperable even from our Western traditions could enable geography itself to perform as leaven in the mass of contemporary science and humanities. The Renaissance of humanism calls for an ecumenical rather than a separatist spirit; it calls for excellence in special fields as well as a concern for the whole picture. It encourages sensitivity to what the barbarism of our own times might be, and it challenges all to seek ways to heal or overcome it in responsible action fully as much as in elegant rhetoric. Fragmentation of thought and life, built into the social fabric of contemporary university curricula, is not overcome by printed appeals to medieval utopias or eulogies on the *uomo universale*. The late twentieth century needs its own Phoenix.

Appendix

Dialogue Project Recordings, 1978–1989

The International Dialogue Project was initiated by Anne Buttimer and Torsten Hägerstrand during the academic year 1977–78 to explore alternative approaches to communication between specialists in the sciences and the humanities, and also between professional experts and the public. Videotaped interviews with senior and retired scholars and professionals who shared insights from their own career experiences, as well as group discussions on specific issues and periods, constituted the core of this process. Recordings were played for audiences from various academic and nonacademic fields as catalysts for dialogue on issues of common concern, as well as evokers of critical reflection on relationships between professional expertise and lived experience. A wide range of individuals, mostly from Europe and America, participated in this process, and over 150 recordings, including some donated by colleagues in other countries, were made during the period 1978–89. Their contents have been analyzed with a view toward discerning common denominators of thought and practice among scholars in various disciplines, as well as the tensions between "internal" and "external" influences in career journeys of twentieth-century scholars (see Chapter 1). These recordings are valuable archival resources for research on knowledge and experience, on theory and practice, and on creativity and context. They are listed here, with reference codes and abbreviated titles, under seven thematically distinct rubrics.

Geography

G 01 Cross-cultural Perspectives on Geography: Aadel Brun-Tschudi (Norway), Olavi Granö (Finland), Wolfgang Hartke (Germany), Torsten Hägerstrand (Sweden), Gerrit Jan van den Berg (The Netherlands), Anne Buttimer (Ireland) 1978
V T (English) PAL Color 51 min.

Note: V = Video recording; A = Audio recording; T = Transcript available; T (GN) = Text published in *Geographers of Norden;* T (PG) = Text published in *The Practice of Geography.* In cases where recordings were made in languages other than English, this is noted in each item. Unless otherwise noted, all tapes and transcripts are available from Media Services at Lund University Library, P.O. Box 3, S-22100 Lund, Sweden.

G 02 The Environment of Graduate School in America: Marvin Mikesell, Leslie Hewes, Preston E. James, Clyde F. Kohn, E. Cotton Mather 1978
V T (PG) NTSC Color 60 min.

G 03 American Geography in the Fifties: George Kish, Duane Knos, Fred Lukermann, Richard L. Morrill, William Pattison 1978
V T (PG) NTSC Color 60 min.

G 04 William William-Olsson 1982
V T PAL Color 56 min.

G 05 Karl Erik Bergsten 1978
V (Swedish) TPAL (English) B&W 43 min.

G 06 Wolfgang Hartke 1979
V (French) T (PG) PAL B&W 45 min.

G 07 T. Walter Freeman 1979
V T PAL Color 38 min.

G 08 Aadel Brun-Tschudi 1979
V T PAL Color 28 min.

G 09 Torsten Hägerstrand 1979
V T PAL B&W 61 min.

G 10 Anne Buttimer 1979
V T PAL B&W 49 min.

G 11 William R. Mead 1979
V T PAL B&W 56 min.

G 12 Sigurdur Thorarinsson 1979
V T (GN) PAL B&W 45 min.

G 13 Axel Sömme 1980
V T (GN) PAL Color I: 58 min.
II: 37 min.

G 14 John Leighly 1980
V T NTSC/PAL B&W 39 min.

G 15 Clarence Glacken 1980
V NTSC/PAL B&W 42 min.

G 16 William J. Talbot 1980
V T PAL Color 34 min.

G 17 Kazimierz Dziewonski 1980
V T PAL B&W 49 min.

G 18 Hans Bobek 1980
V T PAL Color 54 min.

G 19 Michel van Hulten 1980
V T PAL B&W 53 min.

G 20 Robert Geipel 1980
V T NTSC/PAL Color 59 min.

G 21 Jacqueline Beaujeu-Garnier 1981
V T SECAM Color 43 min.

G 22 Peter Scott 1981
V T PAL Color 46 min.

G 23 Robert W. Kates 1981
V T NTSC/PAL Color 55 min.

G 24 On Humanistic Geography: Peter Nash,
Leonard Guelke, Anne Buttimer 1981
V NTSC/PAL Color I: 49 min.
II: 54 min.

G 25 On the Roots of Transformation: Richard Chorley,
Michael Chisholm, Torsten Hägerstrand, Anne Buttimer 1982
V PAL B&W I: 54 min.
II: 33 min.

G 26 Gerrit Jan van den Berg 1982
V PAL Color 57 min.

G 27 Saul B. Cohen 1982
V T NTSC/PAL Color 47 min.

G 28 Fred B. Kniffen 1982
T only

G 29 Speridiao Faissol 1982
V T NTSC/PAL Color 46 min.

G 31 Nikolaj Knattrup 1983
V (Danish) TPAL B&W 53 min.
(GN)

G 32 Derek Gregory 1983
V NTSC/PAL Color 52 min.

G 33 France–USA Encounter: Jacqueline Beaujeu-Garnier, Augustin Berque, Paul Claval, Gérard Dorel, Bernard Marchand, Philippe Pinchemel, discuss with David Hooson, Roger Kasperson, Julian Minghi, Anne Buttimer 1983
V NTSC/PAL Color 55 min.

G 34 The History and Present Condition of Geography. An Historical Materialist Manifesto: David Harvey 1984
V PAL Color 68 min.

G 35 Questions of Integration in Geography. Debate
Held at Queen's University, Belfast 1984
V PAL Color 58 min.

G 36 The Language Prison of Thought and Action 1984
V PAL Color I: 61 min.
II: 21 min.

G 37 Geography in Spain: Juan Vila-Valenti 1984
V (French) PAL Color 56 min.

G 38 Geography in Chile: Hugo Romero 1984
V PAL Color 41 min.

G 39 Patrick Armstrong 1984
V PAL Color 21 min.

G 40 Geography in Japan: Keiichi Takeuchi 1984
V PAL Color 39 min.

G 41 Sitanshu Mookerjee 1984
V PAL Color 38 min.

G 42 The Molding of Geographic Awareness: Peter Nash 1984
V T PAL Color 56 min.

G 43 Geographical Synthesis: J.I.S. Zonneveld 1984
V PAL Color 53 min.

G 44 Environmental Perception in Arid Lands: Les Heathcote 1984
V PAL Color 40 min.

G 45 Geography and Developing Countries: Akin Mabogunje 1984
V PAL Color 43 min.

G 46 Geography in Mexico: Maria Teresa Gutiérrez de MacGregor 1984
V PAL Color 11 min.

G 47 Geography and Social Planning in Poland: Maria Ciechocinska 1984
V PAL Color 38 min.

G 48 Geography, Recreation, and Tourism: J. T. Coppock 1984
V PAL Color 38 min.

G 50 Excursion Printemps 1985: An Excursion by a Group of Swedish Students through the Paris Agglomeration and Brittany 1985
V PAL Color 48 min.

G 51 Jerusalem in the Nineteenth Century: Jehoshua Ben-Arieh 1986
V PAL Color 48 min.

G 52 On the Origins of Academic Geography in Israel: Moise Brawer 1986
V PAL Color 48 min.

G 53 A Window on Geography in Italy Today: Berardo Cori 1986
V PAL Color 21 min.

G 54 A Window on Geography in Ireland: Joseph Haughton 1986
V PAL Color 32 min.

G 55 La Géographie: Est-Elle Utile en Afrique: Kolawole Sikirou Adam 1986
V (French) PAL Color 21 min.

G 56 Géographie et Epistémologie: Daniel Dory 1986
V PAL Color 21 min.

G 57 Geography at the Stockholm School of Economics: Gunnar Alexandersson 1986
V PAL Color 31 min.

G 58 Michael Wise 1986
V PAL Color 30 min.

G 59 Geography in Postwar Poland: Jerzy Kostrowicki 1986
V PAL Color 57 min.

G 60 Paul Claval 1986
V PAL Color 41 min.

G 6l David Hooson			1986
V	PAL	Color	22 min.
G 62 Geography at the Hungarian Academy of Sciences: Marton Pecsi			1986
V	PAL	Color	27 min.
G 63 Karl W. Butzer			1987
V	NTSC	Color	61 min.
G 64 M.R.G. Conzen			1987
V	PAL	Color	60 min.
G 65 Gilbert F. White			1988
V	PAL	Color	30 min.
G 66 Kenneth Hare			1988
V	PAL	Color	47 min.
G 67 Oscar H. K. Spate			1988
V	PAL	Color	45 min.
G 68 Chauncy D. Harris			1988
V	PAL	Color	27 min.

Recordings Donated to Dialogue Project from Karlsruhe[1]

Karlsruher Geovideo No. 5: Ernst Plewe			1984
V (German)	TPAL	Color	60 min.
(English)			
Karlsruher Geovideo No. 6: Hanno Beck			1984
V	PAL	Color	60 min.
Karlsruher Geovideo No. 7: Gottfried Pfeiffer			1985
V	PAL	Color	60 min.

Recordings Donated to Dialogue Project from Barcelona[2]

GGS 1 Antoni F. Tulla			1985
V (Catalan) T	PAL	Color	60 min.
(English)			

1. Copies may be ordered from Geographisches Institut II, Universität Karlsruhe, Kaiserstrasse 12, D-7500 Karlsruhe.

2. Recordings were made in Catalan and Spanish by Joan Nogué i Font, Maria-Dolors Garcia Ramon, and Mireia Belil. English translations of transcripts are being prepared by Joan Nogué i Font, Departamento de Geografia, Estudi General de Girona, Pl. Sant Domenec 9, 17004 Girona, Spain. Tel.: Int + 34 972 213300; FAX: Int + 34 972 216406.

GGS 2 Josep Iglésies 1986
V (Catalan) T PAL Color 60 min.
(English)

GGS 3 On Geography and Didactics: Pilar Benejam 1986
V (Catalan) T PAL Color 60 min.
(English)

GGS 4 Josefina Goméz Mendoza 1986
V (Spanish) T PAL Color 60 min.
(English)

GGS 5 Joaquin Bosque Maurel 1986
V (Spanish) T PAL Color 60 min.
(English)

GGS 6 José Manuel Casas Torres 1986
V (Spanish) T PAL Color 57 min.
(English)

Planning and Development

P 01 Carl-Fredrik Ahlberg 1978
V PAL B&W 55 min.

P 02 The Planning of Stockholm 1978
V PAL B&W I: 56 min.
II: 6 min.

P 03 From Hudiksvall to the Riksdag: Gösta Skoglund 1978
V (Swedish) PAL B&W 55 min.

P 04 Ivar Söderquist 1979
V (Swedish) PAL Color 55 min.

P 05 A Settlement Plan for Israel: Eliezer Brutzkus 1979
V T (English) PAL B&W 32 min.

P 06 Dream and Reality of the Kibbutzim: Chaim Gvati 1979
A T (English)

P 07 Kevin Lynch 1980
V NTSC/PAL Color 57 min.

P 08 Grady Clay — 1981
A T

P 09 Harry Schwarz — 1982
V T — NTSC/PAL — Color — 53 min.

P 10 Göran Sidenbladh — 1982
V — PAL — Color — 58 min.

P 11 Nils Ahrbom — 1982
V — PAL — Color — 60 min.

P 12 Educational Innovation at Worcester Polytechnical Institute — 1981
V — NTSC — Color — 55 min.

P 13 Housing in Singapore: John Humphrey — 1983
V — PAL — Color — I: 45 min.
II: 26 min.

P 14 Manfred Max Néef — 1983
V — PAL — Color — 62 min.

P 15 Nils Stjernquist — 1984
V — PAL — Color — 40 min.

P 16 Lennart Hjelm — 1983
V (Swedish) — PAL — Color — 53 min.

P 17 Olivier Soubeyran — 1986
V (French) — PAL — Color — 43 min.

P 18 Cultural Values and Sudanese Development: Sayyid Hurreiz — 1984
V — PAL — Color — 25 min.

P19 Carin Boalt — 1986
V — PAL — Color — 50 min.

P 20 Jag Maini — 1989
V — NTSC — Color — I: 18 min.
II: 36 min.

P 21 The Gitskan Experience: Neil Sterritt — 1989
V — NTSC — Color — 59 min.

P 22 La Forêt au Québec: Christian Morissonneau — 1989
V (French) — NTSC — Color — 59 min.

Creativity

C 01 Stevan Dedijer — 1979
V T — PAL — B&W — 41 min.

C 02 Vänskapens Hus — 1979
V — PAL — B&W — 37 min.

C 03 Alva Myrdal — 1979
A T

C 04 An Interview with Ilsa Schütz — 1981
A T

C 05 "How to Get There Without Knowing Where You're Going": Alice Higgins — 1980
V T — NTSC/PAL — Color — 39 min.

C 06 Clara Lundh — 1982
V (Swedish) — PAL — Color — 46 min.

C 07 Leon Rappaport — 1983
V (Swedish) — PAL — Color — 120 min.

C 08 Leopold Kohr — 1983
V — PAL — Color — 45 min.

C 09 Märit Rausing — 1987
V (Swedish) — PAL — Color — 80 min.

Health

H 01 Olle Olsson — 1978
V — PAL — B&W — 44 min.

H 02 Gunnar Lindgren — 1978
V — PAL — B&W — 40 min.

H 03 Åke Nordén — 1978
V — PAL — B&W — 48 min.

H 04 Göran Sterky — 1978
V — PAL — B&W — 42 min.

H 05 Ingrid Leodolter — 1980
V T — PAL — B&W — 40 min.

H 06 Gayle Stephens 1980
V NTSC B&W 50 min.

H 07 Caring for Puerto Rican Families: Daniel Amaral and Lucy Candib 1980
V (Spanish and English) NTSC/PAL Color 24 min.

H 08 Pioneers of Dental Science: Doctors Ernest Newburn and Douglas Bratthall 1988
V T PAL Color 29 min.

Enterprise

E 01 A Down-to-Earth Account of the Life of an Inventor: Dag Romell 1979
V PAL B&W 39 min.

E 02 Milton P. Higgins 1980
V T NTSC/PAL Color 40 min.

E 03 Gadelius KK 1979-81
V PAL Color I: 63 min.
II: 54 min.

E 04 Katsuo Shiina, Gadelius KK 1982
V PAL Color I: 58 min.
II: 29 min.

E 05 Karl G. Hjerpe 1982
V PAL Color 62 min.

E 06 Hans Werthén 1983
V PAL Color 52 min.

E 07 Bengt Dieden 1987
V PAL Color 45 min.

E 08 Harald Nerman 1987
V PAL Color 45 min.

E 09 Anneke Rode 1987
V PAL Color 50 min.

E 10 Gustaf Lekholm 1987
V PAL Color 45 min.

E 11 Founding an Institute of Technology in Lund: Bertram Broberg and Olle Härlin
1987
V PAL Color 50 min.

Intercultural Communication

IC 01 Elisabeth Andréasson 1983
V PAL Color 42 min.

IC 02 Being a Stranger: Students in a Foreign Land 1983
V NTSC/PAL Color 54 min.

IC 03 Theodore von Laue 1982
V NTSC/PAL Color I: 58 min.
II: 46 min.

Philosophy and Science

PS 01 Parts and Wholes I: Biology and Language. Marjorie Grene, Hans-Rainer Duncker, Kevin Mulligan, Anne Buttimer 1983
V PAL Color 54 min.

PS 02 Parts and Wholes II: Mathematics and Physics. Ulf Grenander, Abner Shimony, Uno Svedin 1983
V PAL Color 37 min.

PS 03 Parts and Wholes III: Humanities and Social Sciences. Mihailo Markovic, Joseph Agassi, F. Joseph Smith, Anne Buttimer 1983
V PAL Color 49 min.

PS 04 Parts and Wholes IV: Brain Research. Gerhard Roth, Karl Pribram, Uno Svedin
1983
V PAL Color 32 min.

PS 05 Torgny Segerstedt 1984
V (Swedish) PAL Color 63 min.

PS 06 Reflections on Knowledge and Being: André Mercier 1984
V PAL Color 51 min.

PS 07 Islam, Science, and Development: Ibrahim Ahmed Omer 1984
V PAL Color 43 min.

PS 08 Paul Rabinow 1986
V PAL Color 50 min.

PS 09 Som Fysiker Under 20- Och 30-Talet: Torsten Gustafsson 1986
V PAL Color 51 min.

PS 10 A Merchant of Light: Reginald V. Jones 1986
V PAL Color 50 min.

PS 11 The Emancipatory Challenge of Critical Theory: Rick Roderick 1987
V NTSC Color 33 min.

Notes

Introduction

1. The term *human* derives from the Indo-European root *(Dh)ghem-* of the Latin word *humus* = earth. Other derivatives include the Greek term *khthon* = earth; Russian *zemlya* = land; Persian *zamin* = earth, land. The suffixed form *(Dh)ghem-on* means "earthling" (Latin, *humanus*). It is in this etymological sense of "earth dweller" that the term *human* is used throughout this book (Heidegger 1971).

2. See Heidegger 1954, 1971; Buttimer 1976; Seamon and Mugerauer 1985.

3. Two background papers explained the rationale to participants in this venture (Buttimer 1978, 1979; see also Buttimer 1983a).

4. An interim report on the process was presented in Buttimer and Hägerstrand 1980, and a summary presentation of the process, plus vignettes describing interviews conducted between 1978 and 1985, was produced in 1986 (Buttimer 1986). A full list of recordings, now deposited at the Lund University Library, is provided in the Appendix.

5. This trilogy of themes worked well in an attempt to evaluate the "dream and reality" of applied geography at the IGU Congress in Rio de Janeiro (1982). It offered a holistic perspective on water symbolism for hydrologists and theologians at Lund (Buttimer 1984a). It was the framework within which I designed my contribution to a symposium on population policy in the island Pacific in 1983, where I endeavored to transpose the challenge of "migration and identity" to one of potential dialogue and mutual understanding in the encounter between hosts and guests in migrant situations (Buttimer 1985a). And it also worked very well in an attempt to shed light on successive waves of "perception" research in geography for the Geographentag at Münster in 1983 (Buttimer 1984b), as well as for Session 20 of the IGU Congress (Paris 1984), "Trends of Thought and Ideology in Twentieth Century Geography" (Journaux 1985, 110-12).

6. The metaphors discernible in several career accounts were those of journey rather than destination, of groping toward solutions rather than ready-made answers: images of rivers, trees, wanderings in the woods, recurred in several accounts (see Chapter 1).

7. In a forthcoming volume, whose working title is "By Northern Lights," I illustrate how these two interpretive trilogies—meaning, metaphor, milieu, and Phoenix, Faust, Narcissus—could together shed light on geographic thought and practice in Sweden. This case study will be referred to as NL throughout this book.

8. A special issue of *Geojournal* (vol. 26, no. 2, 1992) contains essays on the history of geographical thought in diverse cultural worlds such as ancient China, Japan,

India, Spain, New England, and the Arctic. Other essays in the volume highlight tensions between "academic," "official," and "folk" geographies in colonial situations such as eighteenth-century Guatemala, nineteenth-century Australia, and twentieth-century Latvia.

Chapter One. Meaning, Metaphor, Milieu

Primary sources for this chapter have been the interviews and essays contributed to the International Dialogue Project between 1978 and 1989. Recorded interviews are referenced with their serial number in the list contained in the Appendix. In cases where transcripts have been prepared, page numbers are provided. Essays included in *The Practice of Geography* (1983a) will be referenced with PG, those in *Creativity and Context* (1983b) with CC, and those in *Geographers of Norden* (1988) with GN.

1. Reviewers of *The Practice of Geography* (Buttimer 1983a) offered varied opinions, some rejoicing in the details of individual life experiences, others uneasy about the lack of "explanation." Volumes could be written on the environmental facets of life journeys. Debates and conflicts among "intentionalists" and "structuralists" fill the pages of intellectual history journals (Lilley 1953; Bernal 1965; Mendelsohn et al. 1977; Elzinga 1980). Even more voluminous today are attempts to transcend this antinomy via discourse on language, knowledge, and power (Wittgenstein 1969; Giddens 1979; Foucault 1980; Gregory 1985; Wolch and Dear 1988). What autobiographical accounts afford, in an era of ideological and theoretical controversy, is fresh material for reflection on the manifold and complex interactions between individual and world.

2. Interpretive exercises were conducted in the following manner: At an initial phase, videotaped interviews were shown to audiences at Lund and elsewhere in Sweden, and discussions were held about their content. Later, some videotapes were shown to graduate seminars at Clark University, and students were invited to write down their own personal comments and queries before engaging in discussion. Summaries of the questions which emerged from these discussions were then communicated (in writing) back to the interviewee, who was asked to respond directly to the queries posed or to write an autobiographical essay bearing our queries in mind. Several chose both. From this process, texts were produced which could complement those already available in books and articles. Over a seven-year period, in courses and seminars held in America, France, Ireland, and Sweden, the same general pedagogical procedure was followed in the interpretation of these texts. Each student was assigned a particular life journey for closer scrutiny and was required, not only to study the autobiography and ten selected publications by that author, but also to scrutinize the contexts in which that author's career journey unfolded. At the end of the seminar, each was expected to present an interpretation of a scholar's thought and practice as well as to write an essay on what he or she had derived from the exercise by way of insight into his or her own choice of career.

3. Illustrations of ways in which this interpretive framework could elucidate geographic thought and practice in Sweden were originally designed as part of this book. They are to be contained in a companion volume (see the Introduction).

4. Of course, the sample was a limited one, virtually all from Euro-American

schools. The original aim, however, was to develop a set of themes which could eventually allow for cross-cultural and comparative study. The proposed themes should therefore be regarded as generic, even though they are defined and illustrated here in more restricted ways.

5. The Latin expression *poesis* used throughout this book should be understood as equivalent to the Greek term *po(i)ēsis* (πσ, "a making," "a forming," "a creating").

6. This and the following profiles (figs. 4, 7, 8) should not be taken as typical characterizations of geography in Sweden as a whole. Narrowing the sample to docents meant the exclusion of several notable contributors to Swedish geography and also the later works of those same individuals. Throughout the century, however, docents held responsibility for teaching university courses and may therefore have had the most direct influence on images of geography projected to successive generations of students. NL contains further details of this study and the context within which these and other profiles might be appropriately interpreted.

7. The interpretation of texts was undertaken in three separate rounds. First, all dissertations and reviews of them were read independently by two research assistants, who analyzed contents in terms of meaning, metaphor, milieu, and horizon. Second, the entire exercise was repeated with doctoral students and colleagues in the course of a doctoral seminar held at the University of Lund during autumn 1986. Finally, between 1984 and 1986, colleagues in the departments of geography at Göteborg, Uppsala, and Stockholm were invited to comment critically on the profiles, and their responses shed much light on causes underlying major changes in disciplinary practice.

8. Judgments were made about the relative prominence of four selected categories within each theme. These judgments were then quantized so that the prominence of each category within a theme could be obtained for each dissertation and expressed as a percentage value. Results were contracted so that the prominence of each category within a theme was derived as a percentage value for all dissertations within a given decade. Naturally, all dissertations were given the same weight; that is, the percentage value expressing the prominence of a category for all dissertations within a decade was obtained by adding the individual percentage values for this category and then dividing the result by the number of dissertations. Since only three dissertations within this sample were produced before 1910, they were combined, even though they span a period of more than one decade.

9. Details of this approach are elaborated in the Introduction to Part 2. For an earlier attempt to transpose this schema to geographic thought and practice, see Buttimer 1982.

10. Standard works that illustrate this root metaphor in geography are the programmatic statements of De Geer (1923), Hettner (1927), Hartshorne (1939), Sauer (1941), de Jong (1955), Bunge (1962), Dickinson (1969), and Sack (1980b). Recent research on creativity has reasserted the enduring value of formal metaphors in the genesis of new ideas, be they in art, physics, literature, or mathematics (Mandelbrot 1982).

11. Classical examples of mechanism may be found in the works of Descartes, Comte, Savonarola, Durkheim, and Marx (S. C. Pepper 1942).

12. Alternatively, "organic whole." Pepper's hesitation about the term *organism* is understandable in the light of Anglo-American and Saxon uses of it. In the light of

geography's multilingual and multicultural nature, however, the metaphor of organism seems indeed justifiable and will be used interchangeably with organicism throughout this book.

13. Geographic illustrations of this root metaphor may be found in von Humboldt's *Cosmos,* Vidal de la Blache's work on population, or Semple's *Geography of the Mediterranean Region*. In the neighboring fields of architecture and planning, one finds organismic conceptions expressed, for example, in Park, Burgess, and McKenzie's essays in *The City,* Ebenezer Howard's *Garden Cities for Tomorrow* (1897), Mumford's *Technics and Civilization* (1934), and, more recently, in Prigogine's self-regulating systems (Prigogine and Stengers 1979), holistic medicine (Dossey 1985), and the Gaia Hypothesis (Lovelock 1979).

14. Illustrations of the arena metaphor can best be found in the case studies of social historians such as those of the *Annales* school, in the writings of Febvre and Bloch (Febvre 1922; Bloch 1931), in various urban community studies at Chicago during the 1920s (Park et al. 1925), and most vividly in Jane Jacobs's descriptions of Hudson Street (1961) and later in her analyses of urbanism and regional development (1984). A conceptual rationale for this contextualist approach can be found in the writings of Alfred Schütz and George Herbert Mead (Schütz 1962, 1973). In fact, as a recent essay suggests, here is a style that seems to characterize much present-day humanistic geography (S. Smith 1984).

15. See notes 7, 8.

16. Diverse attempts to design visual representations of career experiences have certainly served well for *paideia.* Students who explore the career profiles of senior scholars readily recognize analogies with their own. A twenty-year-old student in the 1980s, for example, found to his amazement that someone fifty years before had faced challenges not unlike those confronting his generation. The ingenious way in which William-Olsson faced the challenges of the 1930s in Sweden gave hope and inspiration to the career discernment quest of a younger compatriot.

17. The tree itself suggests organic flow: branches suggest fresh starts for ideas or projects; the growing trunk itself is amenable to influences from its changing environment. Some branches may be truncated, depending on an author's energy or the receptivity to his/her ideas among peers or employers; others may push forth buds and fruits of diverse kinds.

18. A cross-section of the rings in the trunk at any particular moment, for example, at moments of career shift, migration, or new scientific challenge, could unmask those various aspects of context and frame of reference which a contextualist would enjoy unraveling. Like the organicist, he or she would rejoice at achieving some kind of synthetic explanation—the organicist for a career journey as a whole, the contextualist for particular career events.

19. The formist might enjoy the classification and mapping of all relevant and tangibly observable indicators of an author's intellectual and practical orientations: carefully documenting, for example, simultaneously occurring orientations in thought and practice (metaphor and meaning) at successive observation points in a career journey. The mechanist would seek connections, possibly of a causal kind, between internal and external forces assumed to have been salient in the production of particular texts. He or she would seek to evoke curiosity about the "ecology" or systems-dynamics of particular career phases or products, and particularly about paradigm change.

20. Historians, for example, tend to prefer orthogonal axes, within which dated information can be presented with time on the vertical axis (up or down). Economists, on the other hand, show a preference for depicting time on the x-axis. In other cases, time is symbolized in terms of a circle or helix.
21. See notes 7, 8.
22. See notes 7, 8.
23. Analogues abound in other fields, such as medical practice. See Sterky (H4) and Stephens (H6) in the International Dialogue Project series (Appendix).

Chapter Two. The Drama of Western Humanism

The substance of this chapter has been published in an article entitled "Geography, Humanism, and Global Concern," *Annals of the Association of American Geographers* 80:1 (1990): 1–33.

1. Lucifer was banished from heaven and Adam from paradise, but the New Adam redeemed humanity; Christ's death by crucifixion was followed by the Resurrection.
2. There are, of course, various versions of this basically Promethean myth. Marlowe's *Dr. Faustus* projected a far more fatalistic outcome than Goethe's *Faust*. Both could find echoes in the history of geographic thought and practice. Elements of the Marlowe version are surely discernible in some contemporary pronouncements about environmental crises. The Goethe version is more open-ended and more inspiring for Western humanity today. Even more inspiring is Shelley's *Prometheus Unbound*.
3. In such a broad-ranging sketch, of course, important nuances within and between different periods of Western history are inevitably glossed over: a mythopoeic account should not be judged in literal terms.
4. Educated in law and Aristotelian philosophy, and fluent in Arabic, Aramaic, and Hebrew, Pico employed kabbalistic theories to elucidate Christian theology. Thirteen of the theses contained in the *Oratio* (1486) were declared heretical by a papal commission. Pico took refuge in the Platonic Academy in Florence and was absolved of heresy by Pope Alexander VI in 1493.
5. "Erasmus is an enemy of all religion," Luther inveighed against the *Manual of the Christian Soldier;* "he is the true adversary of Christ, a perfect replica of Epicurus and Lucian. Whenever I pray, I pray for a curse on him" (cited in de Santillana 1956, 27).
6. Mallarmé thus reawakened the Heraclitean notion of a world in perpetual, unpredictable, and potentially meaningless motion. In stark contrast, Einstein's well-known phrase "God does not play dice" used a similar metaphor to convey his conviction about an absolutely meaningful, that is, eventually explainable universe. These two opposing positions on the prospects for human reason and truth are often overlooked in discussions about relativity and relativism.
7. Take the events of May 1968, a phenomenon that held dramatic significance for virtually all Euro-American students of that generation. Revolutionaries stormed the barricades—Maoists, existentialists, Marxists, and students of law with these and other philosophical inclinations—Phoenix-like in their enthusiasm about human freedom and social engagement, while down the street, in the same city, philosophers with an equally Phoenix-like sense of revolution were busily dismantling (deconstructing) the very notion of human subjectivity. Although activists for social

revolution might discuss Sartre, Mao, even Althusser (Cohn-Bendit and Duteuil 1968), they were apparently nonchalant about that profound critique of individualism, freedom, and active struggle which was then being articulated by the disciples of Lévi-Strauss, Heidegger, and Foucault (Foucault 1966; Aron 1967; Dufrenne 1968; Derrida 1972; *L'Arc* 1980). French philosophy of the 1960s (interpreted by some as simply the transposition to Latin culture of Germanic texts from Marx, Nietzsche, and Freud) cultivated perspectives on humanity and the world which were diametrically opposed to those sung by the ringleaders of May 1968. But could one claim that both types of Phoenix (*logos* and *ergon*) became absorbed or transposed to two opposing Faustian types of development: one to promote the primacy of human agency, the other to promote that of structure? Despite brave attempts to transcend this dualism, from Althusser on, via transactional analysis, structuration theory, semiotics, and other perspectives, the basic tension endures.

8. Rimbaud's (1873) *cri de coeur* stands as frontispiece for André Glucksmann's *Les maîtres penseurs* (1977): "Oh, la science! On a tout repris. Pour le corps et pour l'âme,—le viatique—on a la médecine et la philosophie,—les remèdes de bonnes femmes et les chansons populaires arrangés. Et les divertissements des Princes et les jeux qu'il interdisaient! Géographie, cosmographie, mécanique, chimie! . . . La science, la nouvelle noblesse! Le progrès. Le monde marche! Pourquoi ne tournerait-il pas?" (Oh, science! Everything is included: viaticum for body and soul--medicine, philosophy, good ladies' remedies and arranged popular songs. The recreational diversions of princes and the games they used to forbid. Geography, cosmography, mechanics, chemistry! . . . Science, the new nobility! Progress. The world's a-moving. Why should it not turn around?)

Introduction to Part 2. The Way of Metaphor

1. Metaphor (from *meta-,* "over" or "to change," plus *pherein,* "to carry") states an analogy or a relation between things. It does not imply that things are identical but rather that they are similar in their relations to something else. Strictly speaking, there are various forms of metaphor. *Metonymy* (from *meta-* "to change," plus *onoma* or *onyma,* "name") is the process whereby a name closely associated with an object is used instead of the usual name of the object, for example, "Pentagon," "Ottawa," or "Crown." *Antonomasia* (from *anti-,* "instead of," plus *onoma,* "name") is akin to this: the surname of an individual is used as a generic term for persons of similar orientation or function, for example, Mackintosh, Macadam, Xerox, or Hoover. *Synecdoche* (from *synekdechesthai,* "to receive jointly") is a kind of metaphor whereby a part stands for the whole, for example, "sail" for a ship. In this section, however, the term is used in its most general sense, that is, to cover all these forms of symbolic transformations.

2. The approach sought to overcome the impasse in the relations between science and philosophy which had become so dramatic in the positivist era, sketching a more pluralistic attitude toward knowledge generally.

3. White used the notion of literary trope (metaphor, metonymy, synecdoche, and irony) to capture the author's way of "pre-figuring" the historical field. Within any of these major tropes, he suggested, the author may achieve explanatory effect by combining logical argument, style of narrative, and ideological implication.

Chapter Three. World as Mosaic of Forms

1. Umberto Eco claimed that elements of Porphyry's *Isagoge* continue to affect many contemporary theories of meaning: "The perennial vigor of the Middle Ages is not derived necessarily from religious assumptions, and there is a lot of hidden medievalism in some speculative and systematic approaches of our time, such as structuralism" (1986, 70).

2. Aquinas, a Lombard Norman by birth, and educated by Irish, Danish, and Saxon scholars, was profoundly influenced by Jewish and Arab scholars such as Maimonides and Averröes. He also admired Abelard, the archnominalist, and in fact sought to transcend the gulf separating realism and nominalism. Nor did his conception of the intellectual life (*poesis* and *logos*) exclude concern about practical affairs (*ergon*); he did not hesitate to offer advice to the king of Cyprus about the ideal planning of a city.

3. Traditional interpretations of the *Geographia Generalis,* e.g., those of H. R. Mill (1901), J. Scott Keltie and O. J. R. Howarth (1913), and R. E. Dickinson (1933) all suggested that Special Geography was less important—something to retain student attention—and not quite "scientific" because its data were based on experience and observation rather than on general principles. J.N.L. Baker refuted these interpretations (Baker 1963, 105–18), citing the original work to show that in fact Special Geography was regarded as an integral part of the subject, one "not only most worthy of human study, but one of vital necessity both in the world of letters and for everyday use" (117).

Chapter Four. World as Mechanical System

1. Powerfully attracted to the idea of one universal law that could explain the diversity of earth phenomena, seventeenth-century philosophers hesitated to accept the notion of a deus ex machina. Robert Boyle, seeking to demonstrate that "being addicted to Experimental Philosophy, a Man is rather Assisted than Indisposed, to be a good Christian," offered the following illustration: "First, that a Machine so Immense, so Beautiful, so well contrived, and, in a word, so admirable, as the World, cannot have been the effect of mere Chance, or the Tumultuous Jostlings and Fortuitous Concourse of Atoms, but must have been produc'd by a Cause, exceedingly Powerful, Wise, and Beneficent. Secondly, that this most Potent Author, and (if I may so speak) Opficet of the World, hath not Abandon'd a Masterpiece so worthy of him, but does still Maintain and Preserve it; so regulating the stupendously swift Motions of the great Globes, and other vast Masses of the Mundane Matter, that they do not, by any notable Irregularity, disorder the grand System of the Universe, and reduce it to a kind of *chaos,* or confused State of shuttl'd and depraved things" (Boyle 1690, 29–30). A century later, Goldsmith affirmed that, "when the great Author of Nature began the work of creation, he chose to operate by second causes; and that, suspending the constant exertion of his power, he endued matter with a quality, by which the universal economy of nature might be continued without his immediate assistance. This quality is called *attraction;* a sort of approximating influence, which all bodies, whether terrestrial or celestial, are found to possess; and which in all increases as the quantity of matter in each increases" (Goldsmith 1808, 1:10).

2. In his 1953 address to the Foundation for Integrated Education, John Q. Stewart "likened the body of the Constitution to principles of a rigorous 'field theory' . . . while the Bill of Rights not too fantastically can be said to incorporate corresponding 'quantum conditions.' In politics as well as physics 'the contrast between mass regularity and individual spontaneity is subtle and paradoxical.' The social atom or corpuscle engages in action 'which would be disruptive or impossible for the multitude. But physicists did not find this out until our own times!'" (Cited in Warntz 1964, 156.)

3. Pepper's account of mechanism is based on evidence from before 1940, namely, before the enormous transformations that have occurred in mechanics and engineering.

4. Pepper remarked that this isolation of laws and primary qualities implies a world-view that does not qualify as mechanism—rather, it lapses into formism. "Discrete mechanism is thus internally contradictory" (211).

5. The information needed to predict the writing of these lines was not available at the time of the Big Bang!

6. Latter twentieth-century geographers who have studied environmental perception and cognition are well aware of this. One has sought to identify and correlate "mental maps" and images with patterns of the physical environment, seeking primary explanation either in the perceiver's attributes, intentions, or social reference, on the one hand, or in physical features of the environment itself, on the other (Kates and Wohlwill 1966). Whether one leans toward behaviorist or intentionalist approaches, whether or not one uses expressions such as "perceptual process" or "system," there is as yet little evidence that mechanism can offer a satisfactory account of environmental perception (Saarinen et al. 1984).

7. The relative strength of Davisian views on geomorphology as compared with those of Gilbert in early twentieth-century American geography, as Dorothy Sack pointed out, is not explainable on purely epistemological grounds (Chorley-Beckinsale's hypothesis about the contrast of "closed-" versus "open-systems" cognitive stances). Her interpretation strengthens my hypothesis that in the *foundational* years of discipline building, *poesis* and *paideia* were far more meaningful to audiences and sponsors; *logos* and *ergon* were to become more meaningful in the Faustian phase of the disciplinary story. Gilbert was perhaps a generation or so ahead of his time for general audience appreciation of his analytical and practical art (D. Sack 1991).

8. Long before Darwin's *Origin of Species* (1859), the competing goals of authority and certainty had played on the horizons of European scientists. The record of the late seventeenth and early eighteenth centuries actually reveals the debt that science owes to an attitude that John Kirtland Wright later called *geopiety* and the "theology of nature," which supported scientific inquiry especially in Reformed lands. From Burnet's *Sacred Theory of the Earth* (1684) and John Ray's *Wisdom of God Manifested in the Works of the Creation* ([1690] 1759), through Hutton's *Theory of the Earth* (1795), to Arnold Guyot's *Earth and Man* (1856), the poetic power of religious faith to generate ideas for scientific ways of understanding Planet Earth was surely demonstrated.

9. World War I was to witness the dissolution or demoralization of long-established order within many of these imperial domains, a crumbling of world-view analogous in many ways to those of late medieval times. Evening hours of

empire are prototypically those of contextualist turning, of critical reflection, often of a Narcissus longing for a new Phoenix.

10. An elucidation of meaning, metaphor, milieu in the emergence of early postwar American geography has been attempted in PG, 186-95.

11. Morphogenetic fields can be regarded as analogous to the known fields of physics in that they are capable of ordering physical changes, even though they themselves cannot be observed directly. Gravitational and electromagnetic fields are spatial structures that are invisible, intangible, inaudible, tasteless, and odorless; they are detectable only through their respective gravitational and electromagnetic effects. In order to account for the fact that physical systems influence each other at a distance without any apparent material connection between them, these hypothetical fields are endowed with the property of traversing empty space, or even actually constituting it. In one sense, they are nonmaterial; but in another sense, they are aspects of matter because they can only be known through their effects on material systems. In effect, the scientific definition of matter has simply been widened to take them into account. Similarly, morphogenetic fields are spatial structures detectable only through their morphogenetic effects on material systems; they too can be regarded as aspects of matter if the definition of matter is widened still further to include them (Sheldrake 1981, 72).

12. Is this the Promethean challenge of the late twentieth century: eventually to confound the Lucretian dictum of two millennia ago, "nothing can be produced from nothing," as well as its echo in the first and second laws of thermodynamics?

Chapter Five. World as Organic Whole

1. As Glacken noted, too, "The commonest complaints of the Greeks, chest troubles and malaria, gave evidence of these humors: phlegm, blood (haemorrhaging in fevers), yellow bile, and black bile (the vomiting in remittent malaria)" (Glacken 1967, 11).

2. In paradoxical logic, reality could only be perceived in contradictions; one could not arrive at an understanding of oneness in the cosmos through thought alone. The only way to grasp the world was through the experience of oneness, an experience comparable to the Empedoclean force of love. An altogether different set of ethical standards emerges: the right *way of life* stemmed not necessarily from right *ways of knowing;* rather, it flowed from the experience of oneness. In contrast with Western traditions where major emphasis has been placed on dogma, scientific explanation, and formal codes of morality, the Buddhist tradition sought universal tolerance, emphasis resting on transforming man rather than on transforming nature.

3. "The seeds of destruction lay in the heart of the plant; the worm gnawed at its roots and its vital juices were corrupted: the gigantic tree, therefore, must ultimately fall to the ground" (Herder 1968, bk. 14, chap. 4, 244).

4. The expression "national soul" (*Volkstum*), based on kinship and emotional solidarity, was later to become the focus of harsh criticism, when Herder's ideas were associated with other streams of thought which reified nation-states and expansionary empires. However, as Manuel remarked, "The notion that a thinker should be held morally responsible before some self-appointed Grand Judge for the subsequent fortune of his thought is a patent absurdity" (Herder 1784–91, xvi).

Isaiah Berlin also sought to exonerate Herder from such geopolitical implications, citing remarks such as, "To brag of one's country is the stupidest form of boastfulness. A nation is a wild garden full of bad plants and good, vices and follies mingle with virtues and merit. . . . An innocent attachment to family, language, one's own city, one's own country, its traditions, is not to be condemned." Aggressive nationalism was, however, detestable, and wars criminal. "All large wars are essentially civil wars, since men are brothers, and wars are a form of abominable fratricide" (Berlin 1976).

5. Le Play's comprehensive surveys, *Les ouvriers européens* (1829–51), construed daily life situations in terms of the trilogy of themes *lieu-travail-famille* (place-work-social organization)—an organic conception that was later to inspire Geddes' "Valley Section" (Geddes 1915), Ebenezer Howard's *Garden Cities for Tomorrow* (Howard [1897] 1951), Zimmermann's rural sociology (Zimmerman and Frampton 1935), and Mumford's approach to the history of technology and civilization (Mumford 1934).

6. Not all this environmentalism could be labeled organicist. Strictly speaking, organism and organic models of society found a far more enthusiastic welcome among sociologists than they did among geographers. In the nineteenth century, few metaphors were "so striking or so compelling as the image of the social organism, of society as a living, self-perpetuating, integral and adaptable totality" (Coleman 1966). Geographers seemed content to leave such questions to the rural sociologists or architects; their preference was for the wide open spaces. Besides, it may not have been easy to find in North America the kind of *pays, heimat,* or *seter* which their European colleagues could describe so eloquently with organic metaphors (PG, 62–65).

7. This rejection, and indeed subsequent battles between "creationists" and "evolutionists" in North America, seem odd indeed. Darwin was an exceptionally careful observer of nature, and his theories did not imply a rejection of a Creator. Concluding his volume on the *Origins,* he noted, "Authors of the highest eminence seem to be fully satisfied with the view that each species has been independently created. To my mind it accords better with what we know of the laws impressed on matter by the Creator" (Darwin 1859, 488).

8. It was Daniel Coit Gilman, later to become president of the Johns Hopkins University, who first welcomed Darwinist ideas (Wright 1961). He invited Thomas Huxley to deliver the inauguration address at Hopkins in 1876 and fostered an intellectually open atmosphere where, among others, Frederick Jackson Turner spent some of his formative years. Chamberlain (1843–1928) played a similar role at Chicago.

Chapter Six. World as Arena of Events

1. Organicism also afforded that critical link between an appreciation of cultural relativism and critical self-understanding. At other times, when contexts seemed entirely explicable via systems-analytic models, the contextualist could find common ground with the mechanist: events could then be described, and partially explained, in terms of environmental systems. Biophysiological, economic, or socioculturally defined codes of time and space use, after all, could be construed as "factors," potentially explanatory, coincident or at least coterminous with events. However, when hope stirred that universally recognizable patterns in human perceptions, evalua-

tions, and uses of natural resources could be captured, then formism became a credible route with "mental maps," "culture areas," or whatever the root metaphor of mosaic could afford.

2. This feature might shed light on contextualism's common denominators with existentialism, champion of the human individual's freedom to choose among various alternatives.

3. Peirce actually advanced a linguistic concept of reality, one that was to see further development in the work of Wittgenstein and the neopragmatism of Rorty and others.

4. The life and works of Alfred Schütz offer a treasure of insight on this issue.

5. From the scattered evidence, however, it seems that the late Middle Ages in Europe might have been just the moment when the unconventional was conceivable. Perhaps, as Eco has suggested, Euro-America's postmodernism articulates a nostalgia for the Middle Ages. With his characteristic aplomb, Eco hypothesizes a return to the roots of "Westernism," a cultural style and a *genre de vie* born just in that dark twilight. His argument is a typically contextual one: "events-in-context" then and now are shown to be remarkably analogous (Eco 1986).

6. Images of the world as organism cherish similarly synthetic aims as those of the world as arena. Whereas the former would strive toward generalizations at a global or cosmic scale, however, the latter would seek wholeness in the quality of particular events.

Conclusion. Narcissus Awake

1. The fact that applied geography became such an integral part of national and regional planning in postwar Europe, for example, might be regarded as a case of the Midas touch.

2. Many such emancipatory initiatives, even in the late twentieth century, have been apparently futile, squelched by intransigent authorities or ignored in situations where the majority of people have simply acquiesced in the status quo. Lessons from history suggest that one should never give up hope: that Phoenix may have to die several times before its message is eventually heard.

3. Tensions between the poetic appeal of this metaphor and the overwhelming penchant for reducing GAIA to analytically operational (*logos*) and policy-relevant (*ergon*) terms are today playing out the tensions of Phoenix and Faust.

4. The Gaia Hypothesis certainly refutes the view that life simply adapted to terrestrial conditions as it and they evolved their separate ways (Lovelock 1979, 152).

References

Ackerman, E. A. 1963. Where is a research frontier? *Annals of the Association of American Geographers* 53: 429–40.

Adams, F. D. 1938. *The birth and development of the geological sciences.* London: Baillière Tindall & Cox.

Aeschylus. 1961. *Prometheus Bound, The Suppliants, Seven against Thebes, The Persians.* Trans. Philip Vellacott. Harmondsworth, Middlesex: Penguin Classics.

Ajo, R. 1953. *Contributions to "social physics": A programme sketch with special regard to national planning.* University of Lund, Lund Studies in Human Geography, ser. B, no. 4. Lund: Gleerup.

Alexandersson, G. 1956. *The industrial structure of American cities: A geographic study of urban economy in the United States.* Stockholm: Almqvist & Wiksell.

Althusser, L. 1965. *For Marx.* London: Routledge & Kegan Paul.

Amadeo, D., and R. G. Golledge. 1975. *An introduction to scientific reasoning in geography.* New York: Wiley.

Andrzeyevski, G. 1962. *Ashes and diamonds.* London: Weidenfeld & Nicolson.

Antipode. 1985. The best of Antipode, 1969–1985. *Antipode* 17.

Aristotle. 1941. *The basic works of Aristotle: De Partibus Animalium.* Ed. R. McKeon. New York: Random House.

Aron, A. 1967. *Les étapes de la pensée sociologique.* Paris: Gallimard.

Atkinson, J. L. 1905. The ten Buddhistic virtues (Juzen Hogo): A sermon preached in 1773 by Katsuragi Jiun. *Transactions of the Asiatic Society of Japan* 33: 2.

Augustine, Saint. [418–20] 1972. *The City of God.* Trans. H. Bettenson. Harmondsworth, Middlesex: Penguin Books.

Bachelard, G. [1958] 1964. *La poétique de l'espace.* Paris: Presses universitaires de France. Trans. M. Jolas as *The poetics of space.* New York: Orion Press.

Bacon, F. [1620] 1863–72. *Novum Organum.* In *The works of Francis Bacon,* ed. J. Spedding, R. L. Ellis, and D. D. Heath, 8: 38–69. New York: Herd & Houghton.

Bailey, C. 1926. *The Greek Atomists and Epicurus.* Oxford: Clarendon Press.

Baker, J.N.L. 1963. *The history of geography.* New York: Barnes & Noble.

Baker, O.E.M. 1917. *Geography of the world's agriculture.* Port Washington, N.Y.: Kennikat Press.

Ballasteros, A. G. 1984. Cambios y permanencias en la distribucion espacial de la poblacion espanola (1970–1981). *Anales de geografia de la Universidad Complutense* 4: 83–110.

Barfield, O. 1952. *Poetic diction.* London: Faber.

Barkan, L. 1975. *Nature's work of art: The human body as image of the world.* New Haven: Yale University Press.
Barrows, H. 1923. Geography as human ecology. *Annals of the Association of American Geographers* 13: 1–14.
Baudry, J. 1989. Interactions between agricultural and ecological systems at the landscape level. *Agriculture Ecosystems and Environment* 27: 119–30.
Baum, G. 1975. *Journeys.* New York: Paulist Press.
Bennet, R. J., and R. J. Chorley. 1978. *Environmental systems: Philosophy, analysis, and control.* London: Methuen.
Berdoulay, V. 1978. The Vidal-Durkheim debate. In *Humanistic geography: Prospects and problems,* ed. D. Ley and M. Samuels, 77–90. Chicago: Maroufa Press.
———. 1981. *La formation de l'école française de géographie.* Paris: Bibliothèque Nationale.
———. 1982. La métaphore organiciste: Contribution à l'étude du langage métaphorique en géographie. *Annales de géographie* 507: 573–86.
Bergson, H. L. 1911. *Creative evolution.* New York: Holt.
Berlin, I. 1976. *Vico and Herder: Two studies in the history of ideas.* London: Hogarth Press.
———. 1980. *Against the current: Selected readings.* Ed. H. Hardy. New York: Viking Press.
Berman, M. 1982. *All that is solid melts into air.* Berkeley: University of California Press.
Bernal, J. 1965. *Science in history.* London: C. A. Watts.
Berque, A. 1982. *Vivre l'espace au Japon.* Paris: Presses universitaires de France.
Berry, B.J.L. 1964. Approaches to regional analysis: A synthesis. *Annals of the Association of American Geographers* 54: 2–11.
———. 1972. Cities as systems within cities. In *The conceptual revolution in geography,* ed. W.K.D. Davies, 312–30. Totowa, N.J.: Rowman & Littlefield.
Bertalanffy, L. von. 1969. *General systems theory: Foundations, development, applications.* New York: Braziller.
Bhaskar, R. 1979. *The possibility of naturalism: A philosophical critique of the contemporary human sciences.* Brighton, England: Harvester Press.
Billinge, M., D. Gregory, and R. Martin, eds. 1984. *Recollections of a revolution: Geography as spatial science.* Cambridge: Cambridge University Press.
Birot, P. 1959. *Précis de géographie physique générale.* Paris: Armand Colin.
Black, M. 1962. *Models and metaphors: Studies in language and philosophy.* Ithaca, N.Y.: Cornell University Press.
Blaut, J. 1970. Geographic models of imperialism. *Antipode* 2: 1–7.
———. 1977. Two views of diffusion. *Annals of the Association of American Geographers* 67: 343–49.
Bloch, M. [1931] 1966. *Les caractères originaux de l'histoire rurale française.* Paris: Belles Lettres. Trans. J. Sondheimer as *French rural history: An essay on its basic characteristics.* Berkeley: University of California Press.
Blofield, J., trans. 1965. *I Ching: The book of change.* London: Allen & Unwin.
Blouet, B., ed. 1981. *The origins of academic geography in the United States.* Hamden, Conn.: Archon Books.
Bobek, H. 1928. *Innsbruck, eine Gebirgstadt, ihr Lebensraum und ihre Erscheinung.* Forschungen zur Deutschen Landes- und Volkskunde.

Bodin, J. 1606. *The six books of a commonweale (the republic)*. London: G. Bishop.
Bonnemaison, J. 1985. Les fondements d'une identité: Territoire, histoire, et société dans l'Archipel de Vanuatu (Melanésie). Paris: Thèse pour le Docteur-ès-Lettres et Sciences Humaines, Université de Paris IV.
Bookchin, M. 1974. *The limits of the city*. New York: Harper & Row.
Borges, J. L. 1977. *The book of sands*. New York: Dutton.
Bourdieu, P. 1977. *Outline of a theory of practice*. Cambridge: Cambridge University Press.
———. 1979. The disenchantment of the world. In Bourdieu, *Algeria 1960*, 1–94. Cambridge: Cambridge University Press.
Bowden, M. J. 1980. The cognitive renaissance in American geography: The intellectual history of a movement. *Organon* 14: 199–204.
Bowden, M., and D. Lowenthal, eds. 1975. *Geographies of the mind*. New York: Oxford University Press.
Boyle, R. 1690. *The Christian virtuoso. That by being addicted to experimental philosophy, a man is rather assisted than indisposed to be a good Christian*. London: Printed by Edw. Jones for John Taylor at the Ship, and John Wyat at the Golden Lion, in St. Paul's Churchyard.
———. 1772. The origin of forms and qualities in works. In *The works of the Honorable Robert Boyle*, ed. T. Birch, 1–112. London: J. & F. Rivington, L. Davis et al.
Braudel, F. 1966. *La Méditerranée et le monde méditerranéen à l'époque de Phillipe II*. Paris: Armand Colin.
Breitbart, M. 1981. Peter Kropotkin, the anarchist geographer. In *Geography, ideology, and social concern*, ed. D. Stoddart, 134–53. Oxford: Basil Blackwell.
Broc, N. 1974. *La géographie des philosophes: Géographes et voyageurs français au XVIIIe siècle*. Paris: Editions Ophyrs.
———. 1986. *La géographie de la Renaissance, 1420–1620*. Paris: Editions du Comité des Travaux historiques et scientifiques.
Bronowski, J. 1960. *The common sense of science*. London: Heinemann.
Brookfield, H. C. 1984. Experience of an outside man. In Billinge et al., 27–38.
Brunhes, J. 1902. *L'irrigation, ses conditions géographiques, ses modes et son organisation dans la péninsule Ibérique et dans l'Afrique du Nord*. Paris: C. Naud.
Brunn, S., and E. Yanarella. 1987. Towards a humanistic political geography. *Studies in Comparative International Development* 22: 3–72.
Buchanan, K. 1968. *Out of Asia*. Sydney, Australia: Sydney University Press.
Bunge, W. 1962. *Theoretical geography*. Lund Studies in Geography, ser. C, no. 1. University of Lund.
———. 1973. The geography of human survival. *Annals of the Association of American Geographers* 63: 275–95.
Bunksé, E. V. 1981. Humboldt and an aesthetic tradition in geography. *Geographical Review* 71: 127–46.
———. 1990. Saint-Exupéry's geography lesson: Art and science in the creation and cultivation of landscape values. *Annals of the Association of American Geographers* 80 (1): 96–108.
Burnet, T. [1684] 1691. *Telluris theoria sacra (Sacred theory of the earth)*. London: R. Norton.
Burrow, J. W. 1966. *Evolution and society: A study in Victorian social theory*. Cambridge: Cambridge University Press.

Buttimer, A. 1971. *Society and milieu in the French geographic tradition*. Chicago: Rand McNally.

———. 1974. *Values in geography*. Commission on College Geography Research Report no. 24. Washington, D.C.: Commission on College Geography.

———. 1976. Grasping the dynamism of lifeworld. *Annals of the Association of American Geographers* 66: 277–92.

———. 1978. *On people, paradigms, and "progress" in geography*. Rapporter och notiser 47. University of Lund, Department of Human Geography. Reprinted in Stoddart 1981, 81–98.

———. 1979. Reason, rationality, and human creativity. *Geografiska annaler* 61 B: 43–49.

———. 1982. Musing on Helicon: Root metaphors and geography. *Geografiska annaler* 64 B: 89–96.

———. 1983a. *The practice of geography*. London: Longmans.

———. 1983b. *Creativity and context*. Report on a symposium at Sigtuna, June 1978. University of Lund, DIA Paper no. 2, Lund Studies in Human Geography, ser. B, no. 50. Lund: Gleerup.

———. 1984a. Water symbolism and the human quest for wholeness. In *Vattnet bär livet - funktioner, föreställningar och symbolik*, ed. R. Castensson, 57–91. University of Linköping, Tema (Department of Water in Environment and Society), Report 5: 6. Linköping, Sweden. Abridged version in Seamon and Mugerauer, 159–90.

———. 1984b. *Ideal und Wirklichkeit in der Angewandten Geographie. Münchener Geographische Hefte* 51 (Special issue).

———. 1985a. Farmers, fisherman, gyspies, guests: Who identifies? *Pacific Viewpoint* 26, no. 1: 280–315.

———. 1985b. Integration in geography: Hydra or chimera? In Guelke, 39–68.

———. 1986. *Life experience as catalyst for cross-disciplinary communication: Adventures in dialogue, 1977–1985*. University of Lund, DIA Paper no. 3.

———. 1987. Edgar Kant, 1902–1978. In *Geographers: Bibliographical studies*, no. 11, ed. T. W. Freeman, 71–82. London: Mansell.

———. 1989. *The wake of Erasmus: Saints, scholars, and* studia *in mediaeval Norden*. Lund Studies in Human Geography, no. 54. Lund: Lund University Press.

———. 1990. Geography, humanism, and global concern. *Annals of the Association of American Geographers* 80: 1–33.

———, ed. 1992. *History of geographical thought*. *Geojournal* 26, no. 2 (Special issue).

Buttimer, A., and T. Hägerstrand. 1980. *Invitation to dialogue*. DIA Paper no. 1. University of Lund.

———, eds. 1988. *Geographers of Norden*. Lund Studies in Human Geography, no. 52. Lund: Lund University Press.

Buttimer, A., and D. Seamon, eds. 1980. *The human experience of space and place*. London: Croom Helm.

Butzer, K. W., ed. 1978. *Dimensions of human geography*. University of Chicago, Department of Geography Research Paper no. 186.

———. 1989. Hartshorne, Hettner, and the *Nature of Geography*. In *Reflections on Richard Hartshorne's "The Nature of Geography,"* ed. N. Entrikin and S. Brunn, 35–52. Washington, D.C.: A.A.G. Occasional Publications.

Capel, H. 1981. Institutionalization of geography and strategies of change. In Stoddart 1981, 37–69.
Capot-Rey, R. 1946. *Géographie de la circulation sur les continents*. Paris: Gallimard.
Carcopino, J. 1960. *Daily life in ancient Rome*. New Haven: Yale University Press.
Cassirer, E. 1946. *Language and myth*. Trans. S. Langer. New York: Dover.
———. 1955. *Philosophy of symbolic forms*. New Haven: Yale University Press.
Castells, M. 1977. *The urban question*. London: Edward Arnold.
Chapman, M., ed. 1985. *Mobility and identity in the Island Pacific. Pacific Viewpoint* 26, no. 1 (Special issue).
Cheney, J. 1989. Postmodern environmental ethics: Ethics as bioregional narrative. *Environmental Ethics* 11: 117–34.
Choay, F. 1981. *La règle et le modèle*. Paris: Editions du Seuil.
Chorley, R. J., and P. Haggett. 1967. *Models in geography*. London: Methuen.
Christaller, W. 1968. Wie ich zu der Theorie der zentralen Orte bekommen bin. *Geographische Zeitschrift* 56: 88–101.
Chung-yuan, Chang. 1963. *Creativity and Taoism: A study of Chinese philosophy, art, and poetry*. New York: Julian Press.
Cicero. 1951. *De natura deorum*. Trans H. Rackam. Loeb Classical Library. Cambridge, Mass.: Harvard University Press.
Claval, P. 1964. *Essai sur l'évolution de la géographie humaine*. Paris: Belles Lettres.
———. P. 1980. *Les mythes fondateurs des sciences sociales*. Paris: Presses universitaires de France.
———. 1984. *Géographie humaine et économique contemporaine*. Paris: Presses universitaires de France.
———. 1986. Les géographes français et le monde méditerranéen. Paper presented at symposium, IGU Commission on the History of Geographic Thought, Barcelona.
Cohn-Bendit, D., and J.-P. Duteuil. 1968. *La révolte étudiante: Les animateurs parlent*. Paris: Editions du Seuil.
Colby, C. C. 1933. Centrifugal and centripetal forces in urban geography. *Annals of the Association of American Geographers* 23: 1–20.
Coleman, W. 1966. Science and symbol in the Turner frontier hypothesis. *American Historical Review* 72: 22–49.
Comte, A. 1830–42. *Cours de philosophie positive*. 5 vols. Paris: Sachelier.
Cook, R. 1974. *The tree of life*. London: Thames & Hudson.
Corbridge, S. 1990. Development studies. *Progress in Human Geography* 14: 391–403.
Cosgrove, D. 1984. *Social formation and symbolic landscape*. Totowa, N.J.: Barnes & Noble.
Cox, H. 1985. Moral reasoning and the humanities. In March et al., 25–34.
Crooke, H. 1615. *Microcosmographica: A description of the body of man*. London: W. Jaggard.
Cumont, F. V. 1959. *Die orientalischen religionen in Raemischen Heidentum*. Stuttgart: Teubner.

Dainville, F. de. 1941. *La géographie des humanistes*. Paris: Beauchesne.
———. 1964. *Le langage des géographes*. Paris: Picard.
Daniels, S. 1985. Arguments for a humanistic geography. In *The future of geography*, ed. R. J. Johnston, 143–58. New York: Methuen.

———. 1988. The political iconography of woodland in later Georgian England. In *The iconography of landscape,* ed. D. Cosgrove and S. Daniels, 43–82. Cambridge: Cambridge University Press.

Darby, H. C. 1947. The theory and practice of geography. Inaugural lecture. London: University Press of Liverpool; Hodder and Stoughton.

Dardel, E. 1952. *L'homme et la terre: Nature de la réalité géographique.* Paris: Presses universitaires de France.

Darian, S. G. 1978. *The Ganges in myth and history.* Honolulu: University of Hawaii Press.

Darwin, C. [1859] 1964. *On the origin of species by means of natural selection, or preservation of favoured races in the struggle for life.* Facsimile of 1st ed. Intro. Ernst Mayr. Cambridge, Mass.: Harvard University Press.

Davids, T.W.R. 1881. *Lectures on the origin and growth of religion as illustrated by some points in the history of Indian Buddhism.* London: Williams & Norgate.

Davidson, D. 1978. What metaphors mean. In *On metaphor,* ed. S. Sachs, 29–46. Chicago: University of Chicago Press.

Davies, G. H. 1968. *The earth in decay: A history of British geomorphology, 1578–1878.* London: Macdonald.

Davies, W.K.D. 1972. *The conceptual revolution in geography.* Totowa, N.J.: Rowman & Littlefield.

Davis, W. M. 1906. An inductive study of the content of geography. *Bulletin of the American Geographical Society* 38: 67–84.

Dear, M. 1986. Postmodernism and planning. *Environment and Planning,* D.: *Society and Space,* 4: 367–84.

Deffontaines, P. 1933. *L'homme et la forêt.* Paris: Gallimard.

———. 1948. Défense et illustration de la géographie humaine. *La revue de géographie et d'ethnologie* 1: 5–13.

De Geer, S. 1908. Befolkningens fördelning på Gotland. *Ymer* 28: 240–53.

———. 1923. On the definition, method, and classification of geography. *Geografiska annaler* 5: 1–37.

de Jong, G. 1955. *Het karakter van de geografische totaliteit.* Groningen: J. B. Wolters.

Demangeon, A. 1927a. *Les îles Brittaniques.* 2 vols. Paris: Armand Colin.

———. 1927b. La géographie de l'habitat rural. *Annales de géographie* 36: 1–23, 97–114.

———. 1940. La géographie psychologique. *Annales de géographie* 49: 134–37.

Dematteis, G. 1985. *Le metafore della terra: La geografia umana tra miro e scienza.* Genoa: Gianfiacomo Feltrinelli.

Denecke, D. 1977. Zur Geschichte der Geographie in Göttingen. In *Georgia Augusta,* 77–80. Reprint. Göttingen: Universität Göttingen.

Derrida, J. 1972. *Marges de la philosophie.* Paris: Editions du Minuit.

de Santillana, G. 1956. *The age of adventure.* New York: New American Library.

Descartes, R. [1637] 1968. *Discourse on method and the meditations.* Harmondsworth, Middlesex: Penguin Books.

Dewey, J. 1925. *Experience and nature.* Chicago: Open Court.

Dickinson, R. E. 1969. *The makers of modern geography.* New York: Praeger.

Diderot, D. 1818. *Oeuvres.* Paris: A. Belin.

Diderot, D., and J. R. d'Alembert, eds. 1751–80. *Encyclopédie, ou dictionnaire raisonné des sciences, des arts et des métiers.* 21 vols. Paris: chez Briasson, David l'ainé, Le

Breton Durand; Neuchâtel: chez Samuel Faulèbe & Co. Facsimile 1967. Stuttgart–Bad Cannstatt: Frierich Frommann Verlag (Günther Holzboog).

Dijksterhuis, E. J. 1961. *The mechanization of the world picture.* New York: Oxford University Press.

Dilthey, W. [1913] 1967. *Gesammelte Schriften.* 14 vols. Göttingen: Vandenhoeck & Ruprecht. Vols. 1–12 reissued 1958, Stuttgart: B. G. Teubner.

Doctorow, E. L. 1977. False documents. *American Review* 29: 231–32.

Dossey, L. 1985. *Space, time, and medicine.* Boston: Shambhala, New Science Library.

Doughty, R. 1981. Environmental theology. *Progress in Human Geography* 5: 234–48.

Douglas, M., ed. 1973. *Rules and meanings.* Harmondsworth, Middlesex: Penguin Books.

Downs, R. M., and D. Stea, eds. 1973. *Image and environment: Cognitive mapping and spatial behavior.* Chicago: Aldine.

Dreyfus, H., and P. Rabinow. 1984. *Un parcours philosophique.* Paris: Gallimard.

Dryer, C. R. 1920. Genetic geography: The development of geographic sense and concept. *Annals of the Association of American Geographers* 10: 3–16.

Dubos, R. 1972. *Man adapting.* New Haven: Yale University Press.

Dufrenne, H. 1968. *Pour l'homme.* Paris: Editions du Seuil.

Dunbar, G. S. 1981. Elisée Reclus, an anarchist in geography. In Stoddart 1981, 154–64.

Duncan, E. H. 1951. Satan-Lucifer: Lightning and thunderbolt. *Philological Quarterly* 30: 441–43.

Duncan, J. S. 1980. The superorganic in American cultural geography. *Annals of The Association of American Geographers* 70: 181–98.

Duncker, H.-R. 1985. Organisms, complexity, teleology, and conflicting heuristic principles. In *Parts and wholes,* vol. 3, ed. P. Sällström, 27–40. Stockholm: Swedish Council for Planning and Coordination of Research, Committee for Future Oriented Research.

Eco, U. 1986. *Travels in hyperreality.* New York: Harcourt Brace Jovanovich.

Efron, A., and J. Herold, eds. 1980. *Root metaphor: The live thought of Stephen C. Pepper.* Special Issue of *Paunch.* New York: State University of Buffalo, Department of English.

EFTA Economic Development Committee. 1974. *National settlement strategies: A framework for regional development.* Geneva: Secretariat of EFTA.

Ehrenfeld, D. 1978. *The arrogance of humanism.* New York: Oxford University Press.

Einstein, A., and L. Infeld. 1938. *The evolution of physics: The growth of ideas from the early concepts to relativity and quanta.* Cambridge: Cambridge University Press.

Eisel, U. 1976. *Die Entwicklung der Anthropogeographie von einer "Raumwissenschaft" zur Gesellschaftswissenschaft.* Urbs et Regio no. 17. Kassel: Gesamthochschul-Bibliotek.

Eliade, M. 1969. *The quest.* Chicago: University of Chicago Press.

Elzinga, A. 1980. Models in the theory of science: A critique of the convergence thesis. In *Technological change and cultural impact in Asia and Europe: A critical review of the Western theoretical heritage,* ed. E. Baart, A. Elzinga, and B.-E. Borgström, 37–70. Report no. 32-S. Stockholm: Swedish Council for Planning and Coordination of Research, Committee for Future Oriented Research.

Emerson, R. W. 1803–82. *The complete works.* Cambridge, Mass.: Harvard University Press, Belknap Press.
Entrikin, N. 1976. Contemporary humanism in geography. *Annals of the Association of American Geographers* 66: 615–32.
Evans, E. E. 1983. *The personality of Ireland: Habitat, heritage, and history.* Cambridge: Cambridge University Press.

Febvre, L. 1922. *La terre e l'évolution humaine.* Paris: Renaissance du livre.
Ferguson, M. 1982. *The Aquarian conspiracy: Personal and social transformations in the 1980s.* London: Granada Books.
Ferrier, J.-P., J.-P. Racine, and C. Raffestin. 1978. Vers un paradigme critique: Matériaux pour un project géographique. *L'Espace géographique* 4: 291–97.
Ferry, L., and A. Renaut. 1985. *La pensée 68: Essai sur l'anti-humanisme contemporain.* Paris: Gallimard.
Feyerabend, P. K. 1961. *Knowledge without foundations.* Oberlin, Ohio: Oberlin University Press.
———. 1975. *Against method: Outline of an anarchist theory of knowledge.* London: New Left Books.
Firey, W. 1960. *Man, mind, and land.* Glencoe, Ill.: Free Press.
Fish, S. 1981. *Is there a text in this class? The authority of interpretive communities.* Cambridge, Mass.: Harvard University Press.
Fleck, L. [1935] 1979. *Entsehung und Entwicklung einer Wissenschaftsche tatsache.* Basel: Verbags Buchandlung. Ed. and trans. F. Bradley and T. J. Trenn as *Genesis and development of a scientific fact.* Chicago: University of Chicago Press.
Folch-Serra, M. 1989. Geography and post-modernism: Linking humanism and development studies. *Canadian Geographer* 33: 66–75.
Forrester, J. W. 1969. *Urban dynamics.* Cambridge, Mass.: MIT Press.
Forster, G. 1777. *A voyage round the world.* London: Printed for B. White, J. Robson, P. Elmsly, and G. Robinson.
Foucault, M. 1966. *Les mots et les choses.* Paris: Gallimard.
———. [1975] 1979. *Surveiller et punir: Naissance de la prison.* Paris: Gallimard. Trans. Alan Sheridan as *Discipline and punish.* Harmondsworth, Middlesex: Penguin Books.
———. 1980. *Power and knowledge.* Trans. C. Gordon. Sussex: Harvester Press.
Frängsmyr, T., ed. 1983. *Linnaeus: The man and his work.* Berkeley: University of California Press.
———. 1989. *Science in Sweden: The Royal Swedish Academy of Sciences, 1739–1989.* Stockholm: Royal Swedish Academy of Sciences.
Fraser, N. 1975. *Of time, passion, and knowledge.* New York: Braziller.
Fremantle, A. 1954. *The age of belief.* New York: New American Library.
Frye, N. 1981. The bridge of language. *Science* 212: 127–32.

Gadamer, H.-G. [1965] 1975. *Wahrheit und Methode.* 2d ed. Tubingen: J. C. Mohr. Ed. and trans. G. Bardin and J. Cumming as *Truth and method.* New York: Seabury Press.
Gale, S., and G. Olsson, eds. 1979. *Philosophy in geography.* Dordrecht, Netherlands: Reidel.

Galtung, J. 1981. Sivilisasjon, kosmologi, fred og utvikling. Oslo: *Det Norske Videnskapsakademi årsbok,* 130–53.

Gardner, H. 1983. *Frames of mind: The theory of multiple intelligences*. New York: Basic Books.

Geddes, P. 1915. *Cities in evolution*. London: Oxford University Press.

Geertz, C. 1983. The way we think now: Toward an ethnography of modern thought. In *Local knowledge: Further essays in interpretive anthropology,* 147–66. New York: Basic Books.

Gibbs-Smith, C. 1985. *The inventions of Leonardo da Vinci*. London: Peerage Books.

Giddens, A. 1979. *Central problems in social theory: Action, structure, and contradiction in social analysis*. London: Macmillan.

Gilgamesh. 1960. *The epic of Gilgamesh*. Ed. N. K. Sanders. Harmondsworth, Middlesex: Penguin Books.

Gilson, E. 1950. *The spirit of mediaeval philosophy*. London: Sheed & Ward.

Glacken, C. 1967. *Traces on the Rhodean shore*. Berkeley: University of California Press.

Glucksmann, A. 1977. *Les maîtres penseurs*. Paris: Grasset.

Godlund, S. 1986. Swedish social geography. In *Social geography in international perspective,* ed. J. Eyles, 96–150. Totowa, N.J.: Barnes & Noble.

Goethe, J. W. von. 1808, 1833. *The tragedy of Faust*. [1909] 1982. *Johann Wolfgang von Goethe and Christopher Marlowe*. Ed. Charles W. Eliot. Harvard Classics. Danbury, Conn.: Grolier.

———. [1790] 1952. *Goethe's botanical writings*. Trans. B. Mueller. Honolulu: University of Hawaii Press.

Goldsmith, O. 1808. *A history of the earth and animated nature*. 4 vols. High-Ousegate, York: Thomas Wilson and Son.

———. 1887. The logicians refuted. In *Poems of Goldsmith and Gray*. New York: Co-operative Publication Society.

Gomez-Mendoza, J., and N. G. Cantero. 1992. Interplay of state and local concern in the management of natural resources: Hydraulics and forestry in Spain, 1835–1936. In A Buttimer, ed., *Geojournal* 26 (2): 173–80.

Gould, P. 1985. *Geographers at work*. London: Routledge & Kegan Paul.

Gould, P., and G. Olsson, 1982. *A search for common ground*. London: Pion.

Gould, S. J. 1987. *Time's arrow, time's cycle: Myth and metaphor in the discovery of geological time*. Cambridge, Mass.: Harvard University Press.

Granö, J. G. 1929. *Reine geographie: Eine methodologische studie beleuchtet mit Beispielen aus Finnland and Estland*. Acta Geographica, 2: 2. Helsinki: Geographical Society of Finland.

Granö, O. 1981. External influences and internal change in development. In Stoddart 1981, 17–36.

Grassi, E. 1979. The priority of common sense and imagination: Vico's philosophical relevance today. In Tagliacozzo et al., 163–87.

Gregory, D. 1978. *Ideology, science, and human geography*. London: Hutchinson.

———. 1981. The ideology of control: Systems theory and geography. In *Tidjschrift voor Economische en Sociale Geografie* 71: 327–41.

———. 1985. Time and space in social life. Clark University, Graduate School of Geography, Atwood Lecture Series no. 1. Worcester, Mass.

Groot, H. de. 1910. *Religion of the Chinese*. New York: N.p.

Guelke, L., ed. 1985. *Geography and humanistic knowledge.* University of Waterloo, Department of Geography, Waterloo Lectures in Geography, 2. Waterloo, Ontario.

Guyot, A. 1856. *The earth and man: Lectures on comparative physical geography in its relation to the history of mankind.* London: Richard Bentley.

Habermas, J. 1968. *Knowledge and human interests.* Boston: Beacon Press.

———. 1976. *Communication and the evolution of society.* Boston: Beacon Press.

Hacker, L. M. 1947. *The shaping of the American tradition.* New York: Columbia University Press.

Hägerstrand, T. 1953. *The propagation of innovation waves.* Lund Studies in Geography, ser. B, no. 4. University of Lund.

———. 1970. What about people in regional science? *Papers of the Regional Science Association* 24: 7–21.

———. 1982. Proclamations about geography from the pioneering years in Sweden. *Geografiska annaler* 64 B: 119–25.

Haggett, P. 1965. *Locational analysis in human geography.* London: Edward Arnold.

———. 1972. *Geography: A modern synthesis.* New York: Harper & Row.

———. 1990. *The geographer's art.* Oxford: Basil Blackwell.

Haig, R. M. 1928. *Regional survey of New York and its environs.* Vol. 1, *Major economic factors in metropolitan growth and arrangement.* New York: McCrea.

Haldane, J.B.S. 1939. The theory of the evolution of dominance. *Journal of Genetics* 37: 365–74.

Halévy, E. 1952. *The growth of philosophical radicalism.* London: Faber & Faber.

Hampshire, S., ed. 1956. *The age of reason: The seventeenth-century philosophers.* New York: New American Library.

Hannerberg, D., ed. 1957. *Migration in Sweden: A symposium.* Lund Studies in Geography, ser. B, no. 3. University of Lund.

Haraway, D. J. 1976. *Crystals, fabrics, and fields.* New Haven: Yale University Press.

Hard, G. 1973. *Die geographie, eine wissenschaftstheoretische Einfhuring.* Berlin: Walter de Gruyter.

Hardy, G. 1939. *La géographie psychologique.* Paris: Gallimard.

Harris, C. 1978. The historical mind and the practice of geography. In Ley and Samuels, 123–37.

Harris, C., and E. L. Ullman. 1945. The nature of cities. *Annals of the American Academy of Political and Social Sciences* 242: 7–17.

Hart, J. F. 1979. The 1950s. *Annals of the Association of American Geographers* 69: 109–14.

Hartley, D. 1749. *Observations on man: His frame, his duty, and his expectations.* London: S. Richardson.

Hartshorne, R. [1936] 1939. The nature of geography: A critical survey of current thought in the light of the past. Lancaster, Pa.: Association of American Geographers. Reprinted in *Annals of the Association of American Geographers* 29 (entire volume).

———. 1959. *Perspective on the nature of geography.* Chicago: Rand McNally.

———. 1964. Robert S. Platt, 1891–1964. *Annals of the Association of American Geographers* 54: 630–37.

Harvey, D. 1969. *Explanation in geography.* New York: St. Martin's Press.

———. 1972. Revolutionary and counter-revolutionary theory in geography and the problem of ghetto formation. *Antipode* 4: 1–12.
———. 1973. *Social justice and the city.* London: Edward Arnold.
———. 1984. On the history and present condition of geography: An historical materialist manifesto. *Professional Geographer* 36: 1–10.
Hasr, S. H. 1964. *An introduction to Islamic cosmological doctrines: Conceptions of nature and methods used for its study by the Ikhwan al Safa, al Biruni, and Ibn Sina.* Cambridge, Mass.: Harvard University Press, Belknap Press.
Hawking, S. 1988. *A brief history of time.* New York: Bantam Books.
Hawthorne, N. [1844] 1972. *Tales and sketches including "Twice-told tales," "Mosses from an old manse," and "The snow image."* Columbus: Ohio State University Press.
Heidegger, M. 1947. *Platons Lehre von der Wahrheit: Mir einen Brief über den "Humanismus."* Bern: A. Francke.
———. 1954. *Aus der Erfahrung des Denkens.* Pfullingen: Neske.
———. 1971. *Poetry, language, and thought.* New York: Harper.
Helmfrid, S. 1959. De geometriska jordböckernas skattläggningskartor. *Ymer* 3: 224–31.
Herbert, D. T., and R. J. Johnston, eds. 1978. *Social areas in cities.* Vol. 1, *Spatial processes and form.* New York: Wiley.
Herbertson, A. J. 1905. The major natural regions: An essay in systematic geography. *Geographical Journal* 25: 300–312.
Herder, J. G. von. [1784–91] 1968. *Ideen zur Philosophie der Geschichte der Menscheit.* (*Reflections on the philosophy of the history of mankind*). Ed. F. E. Manuel. Chicago: University of Chicago Press.
Herodotus. 1987. *The history of Herodotus.* Trans. David Grene. Chicago: University of Chicago Press.
Hesse, M. 1966. *Models and analogies in science.* Notre Dame, Ind.: University of Notre Dame Press.
Hettner, A. 1905. Das Wesen und die Methoden der Geographie. *Geographische Zeitschrift* 11: 545–64, 615–29, 671–86.
———. 1927. *Die Geographie: ihre Geschichte, ihr Wesen, und ihre Methoden.* Breslau: F. Hirt.
Hippocrates. 1931. *Nature of man.* Vol. 4, *Being.* Trans. W.H.S. Jones. Loeb Classical Library. New York: Putnam.
Hocquenghem, G. 1986. *Lettre ouverte à ceux qui sont passés du col Mao au Rotary.* Paris: Albin Michel.
Hofstadter, D. R. 1979. *Gödel Escher Bach: An eternal golden braid.* New York: Basic Books.
Horkheimer, M., and T. Adorno. [1944] 1972. *Dialektik der Auflkärung.* New York: Social Studies Association. Trans. J. Cumming as *Dialectics of enlightenment.* New York: Herder & Herder.
Howard, E. [1897] 1951. *Garden cities for tomorrow.* London: Faber.
Hudson-Rodd, N. 1991. Place and health in Canada: Historical roots of two healing traditions. Ph.D. diss., University of Ottawa, Department of Geography.
Huff, D. L. 1960. A topographical model of consumer space preferences. *Papers and Proceedings of the Regional Science Association* 6: 159–73.
Humboldt, A. von. [1845–62, 1850–59] 1848. *Kosmos: Entwurf einer physischen Welt-*

beschreibung. Trans. E. C. Otté as *Cosmos: Sketch of a physical description of the universe*. 5 vols. London: Henry G. Bohm.

———. 1849. *Aspects of nature in different climates*. Philadelphia: Lea & Blanchard.

Huntington, E. 1915. *Civilization and climate*. New Haven: Yale University Press.

———. 1926. *The human habitat*. 4th ed. New York: Van Nostrand.

———. 1945. *Mainsprings of civilization*. New York: Wiley.

———. 1945. Regions and seasons of mental activity. In *Mainsprings of civilization,* 343–67. New York: Wiley.

Hutton, J. 1795. *Theory of the earth with proofs and illustrations*. Edinburgh: Cadell, Junion, & Davies.

Isard, W., D. F. Bramhall, G.A.P. Carrothers, J. H. Cumberland, L. N. Moses, D. O. Price, and E. W. Schooler, eds. 1960. *Methods of regional analysis: An introduction to regional science*. New York: Wiley.

Jackson, J. B., ed. 1952. Human, all too human geography. *Landscape* 2: 5–7.

———. 1976. The domestication of the garage. *Landscape* 20: 19.

Jacobs, D. 1969. *Master builders of the Middle Ages*. New York: American Heritage.

Jacobs, J. 1961. *The death and life of great American cities*. New York: Random House.

———. 1984. *Cities and the wealth of nations*. Harmondsworth, Middlesex: Penguin Books.

Jager, B. 1975. Theorizing, journeying, dwelling. In *Duquesne Studies in Phenomenological Psychology,* ed. A. Giorgi, W. F. Fischer, and R. von Eckartsberg, 2: 235–60. Pittsburgh: Duquesne University Press.

Jakobson, R., and M. Halle. 1956. *Fundamentals of language*. The Hague: Mouton.

James, P. E. 1981. Geographical ideas in America, 1890–1914. In *The origins of academic geography in the United States,* ed. B. Blouet, 319–26. Hamden, Conn.: Archon Books.

James, P. E., and C. F. Jones. 1954. *American geography: Inventory and prospect*. Syracuse, N.Y.: Syracuse University Press.

James, P. E., and G. Martin. 1979. *Association of American Geographers: Seventy-five years, 1905–1975*. Washington, D.C.: Association of American Geographers.

James, W. [1907] 1955. *Pragmatism and four essays from "The meaning of truth."* New York: New American Library.

Jameson, F. 1983. Postmodernism and consumer society. In *The anti-aesthetic: Essays on postmodern culture,* ed. H. Foster, 111–25. Port Townsend, Wash.: Bay Press.

Janik, A., and S. Toulmin. 1973. *Wittgenstein's Vienna*. London: Weidenfeld & Nicolson.

Jefferson, M. 1928. The civilizing rails. *Economic Geography* 4: 217–32.

Johnston, R. J. 1979. *Geography and geographers*. London: Edward Arnold.

———. 1983. *Philosophy and human geography*. London: Edward Arnold.

Johnston, R. J., and P. Taylor. 1986. *A world in crisis*. Oxford: Basil Blackwell.

Jolly, M. 1982. Birds and banyans of south Pentecost: *Kastom* in the anti-colonial struggle. *Mankind* 13: 338–56.

Jones, M., ed. 1986. *Welfare and environment*. Trondheim, Norway: Tapir Press.

Journaux, M., ed. 1985. *Comptes rendus du XXV Congrès International Géographique*. Paris: Comité National de l'UGI.

Judson, H. F. 1979. *The eighth day of creation: The makers of the revolution in biology.* New York: Simon & Schuster.
———. 1980. *The search for solutions.* New York: Holt, Rinehart, & Winston.
Jung, C. G. 1964. *Man and his symbols.* New York: Doubleday.

Kale, R. M. N.d. *The Meghaduta of Kalidasa.* 6th ed. Bombay: Booksellers.
Kant, E. 1934. *Problems of environment and population in Estonia.* Publicationes Seminarii Universitatis Oeconomico-Geographici, 7. Tartu: University of Tartu.
———. 1935. *Estland und Baltoskandia: Bidrag till Östersjöländernas geografi och sociografi.* Svio-Estonica, 79–103.
———. 1946. Den inre omflyttningen i Estland i samband de Estniska Studernas Omland (Internal migration in Estonia in the context of Estonia's Town Hinterlands). *Svensk geografisk årsbok* 22: 83–124.
———. 1962. Classification and problems of migration. In *Readings in cultural geography,* ed. P. L. Wagner and M. W. Mikesell, 342–54. Chicago: University of Chicago Press.
Kant, I. 1802. *Physische Geographie.* Leipzig: Felix Meiner.
———. [1787] 1855. *Critique of pure reason.* Trans. J.M.D. Meiklejohn. London: Henry G. Bohm.
Karjalainen, P. T. 1986. *Geodiversity as a lived world: On the geography of existence.* University of Joensuu Publications in Social Sciences, no. 7. Joensuu, Finland.
Karjalainen, P. T., and P. Vartainen. 1990. Dwelling in geography. *Nordisk samhallsgeografisk tidskrift* 11: 98–105.
Kates, R. W. 1969. Mirror or monitor for man? *Antipode* 1: 47–53.
———. 1987. The human environment: The road not to take, the road still beckoning. *Annals of the Association of American Geographers* 77: 525–34.
Kates, R. W., and J. Wohlwill, eds. 1966. *Man's response to the physical environment. Journal of Social Issues* 22 (Special issue).
Kay, J. 1989. Human dominion over nature in the Hebrew bible. *Annals of the Association of American Geographers* 79: 214–32.
Kearney, R. 1988. *The wake of imagination.* London: Hutchinson.
Kearns, R. 1988. In the shadow of illness: A social geography of the chronically mentally disabled in Hamilton, Ontario. Ph.D. diss., Department of Geography, McMaster University, Hamilton.
Keat, R., and J. Urry. 1975. *Social theory as science.* London: Routledge & Kegan Paul.
Keller, S. 1983. *The urban neighborhood.* New York: Random House.
Kiliani, M. 1983. *Les cultes du cargo mélanesien: Mythe ete rationnalité en anthropologie.* Lausanne: Editions d'en bas.
Kinami, T., ed. 1973. *Jiun, Juzen Hogo.* Kyoto: Sanmitsudo Shoten.
Kirk, D. 1946. *Europe's population in the interwar years.* New York: Gordon & Breach.
Kirk, G. S. 1974. *The nature of Greek myths.* Harmondsworth, Middlesex: Penguin Books.
Kirk, G. S., and J. E. Raven. 1962. *The pre-Socratic philosophers.* Cambridge: Cambridge University Press.
Kirk, W. 1951. Historical geography and the concept of the behavioral environment. *Indian Geographical Journal,* 152–60 (Silver Jubilee volume).

Koelsch, W. A. 1970. Scholarly issues. Contribution to a symposium on Values in Geography held at Clark University. In Buttimer 1974, 32–33.
———. 1979. Nathaniel Southgate Shaler (1841–1906). In *Geographers: Biobibliographical studies,* ed. P. Pinchemal and T. W. Freeman, 133–39. London: Mansell.
Koestler, A. 1978. *Janus: A summing up*. London: Hutchinson.
Kohak, E. 1984. *The embers and the stars: A philosophical inquiry into the moral sense of nature*. Chicago: University of Chicago Press.
Koyré, A. 1957. *From the closed world to the infinite universe*. Baltimore: Johns Hopkins Press.
———. 1973. *Etudes d'histoire de la pensée scientifique*. Paris: Gallimard.
Kramer, S. N. 1959. *History begins at Sumer.* New York: Doubleday.
Kropotkin, P. 1885. What geography ought to be. *Nineteenth Century* 18: 940–56.
———. 1898. *Fields, factories, and workshops*. London: T. Nelson.
———. [1902] 1914. *Mutual aid: A factor in evolution*. Boston: Extending Horizons Books.
Kuhn, T. S. 1970. *The structure of scientific revolutions*. 2d ed. Chicago: University of Chicago Press.
Kunze, D. 1984. Thought and place: The imagination and memory of eternal places in the philosophy of Giambattista Vico. Ph.D. diss., Department of Geography, Pennsylvania State University, State College.

Lakoff, G., and M. Johnson. 1980. *Metaphors we live by.* Chicago: University of Chicago Press.
Langer, K. S. 1957. *Philosophy in a new key: A study in the symbolism of reason, rite, and art*. Cambridge, Mass.: Harvard University Press.
Laplace, P. S., Marquis de. [1796] 1966. *Celestial mechanics*. 4 vols. Trans. N. Bowditch. New York: Chelsea.
L'Arc. 1980. Gilles Deleuze. No. 49 (Special issue).
Lavedan, G. 1936. *Géographie des villes*. Paris: Gallimard.
Lawson, V., and L. A. Staeheli. 1990. Realism and the practice of geography. *Professional Geographer* 42: 13–19.
Leatherdale, W. H. 1974. *The role of analogy, model, and metaphor in science*. Amsterdam: North Holland.
Leboulaye, E., ed. 1975–79. *Oeuvres complètes de Montesquieu*. Paris: Garnier Frères.
Lebret, L. J. 1961. *Dynamique concrète du dévelopement*. Paris: Editions ouvrières.
Leclercq, J. 1963. *Introduction à la sociologie*. 3d ed. Paris: Beatrice Nauwelaerts.
Lefèbvre, H. 1974. *La production de l'espace*. Paris: Editions Anthropos.
Leighly, J. 1955. What has happened to physical geography? *Annals of the Association of American Geographers* 45: 309–18.
Leiss, W. 1974. *The domination of nature*. Boston: Beacon Press.
Lenoble, R. 1969. *Esquisse d'une histoire de l'idée de nature*. Paris: Albin Michel.
Le Play, F. 1855. *Les ouvriers européens*. Tours: Alfred Mame.
Lévi-Strauss, C. 1966. *The savage mind*. London: Weidenfeld & Nicolson.
Lewin, K. 1952. *Field theory in social science*. London: Tavistock.
Lewis, P. F. 1979. Axioms for reading the landscape. In *The interpretation of ordinary landscapes,* ed. D. Meinig, 11–32. New York: Oxford University Press.

———. 1986. Beyond description. *Annals of the Association of American Geographers* 76: 465–78.

Lewis, R.W.B. 1955. *The American Adam*. Chicago: University of Chicago Press.

Ley, D. 1980. Geography without man: A humanistic critique. Occasional paper. Oxford University, School of Geography.

———. 1981, 1983. Cultural/humanistic geography. *Progress in Human Geography* 5: 249–57; 7: 267–75.

———. 1989. Fragmentation, coherence, and limits to theory in human geography. In Kobayashi and Mackenzie, 223–44.

Ley, D., and M. Samuels, eds. 1978. *Humanistic geography: Prospects and problems*. Chicago: Maroufa Press.

Lidmar-Bergström, K. 1984. Geografi, geomorfologi och geovetenskap i skolan, vid universitet och i samhället. *Geografiska notiser* 42: 23–27.

Light, R. U. 1944. Progress in medical geography. *Geographical Review* 34: 632–41.

Lilley, S. 1953. Cause and effect in the history of science. *Centaurus* 3: 58–72.

Lindqvist, S. 1984. *Technology on trial: The introduction of steam power technology into Sweden, 1715–1736*. Uppsala Studies in the History of Science, no. 1. Stockholm: Almqvist & Wiksell.

Linnaeus, C. [1734] 1980. *Iter Dalkarlicum*. In *Linné i Dalarna*, ed. B. Gullander. Stockholm: Geber.

Livingstone, D. 1982. Nature, man, and God in the geography of Nathaniel S. Shaler. Ph.D. diss., Department of Geography, Queens University, Belfast.

Livingstone, D., and R. Harrison. 1981. Meaning through metaphor: Analogy as epistemology. *Annals of the Association of American Geographers* 71: 95–107.

Løffler, E. 1911. *Min selvbiographi: En geographs levnadslb*. Copenhagen: Lehmann & Stages Boghandel.

Lopez, B. 1986. *Arctic dreams: Imagination and desire in a northern landscape*. London: Macmillan.

Lösch, A. 1954. *The economics of location*. New Haven: Yale University Press.

Lovejoy, A. O. 1936. *The great chain of being: A study of the history of an idea*. Cambridge, Mass.: Harvard University Press.

———. 1961. *Reflections on human nature*. Baltimore: Johns Hopkins Press.

Lovelock, J. E. 1979. *Gaia: A new look at life on earth*. New York: Oxford University Press.

Lowenthal, D. 1961. Geography, experience, and imagination: Towards a geographic epistemology. *Annals of the Association of American Geographers* 51: 241–60.

Lowenthal, D., ed. 1967. *Environmental perception and behavior*. University of Chicago, Department of Geography.

Lubac, H. de. 1944. *Le drame de l'humanisme athée*. Paris: Spes.

Lucretius. [1478] 1982. *De rerum natura*. Trans. W.H.D. Rouse and M. F. Smith as *The nature of things*. Loeb Classical Library. Cambridge, Mass.: Harvard University Press; London: W. Heinemann.

Lukerman, F. 1965. The "calcul des probabilités" and the école française de géographie. *Canadian Geographer* 9: 128–37.

Lyell, C. 1830–33. *Principles of geology, being an attempt to explain the former changes of the earth's surface by reference to causes now in operation*. London: John Murray.

Lynch, J. J. 1985. *The language of the heart: The body's response to human dialogue*. New York: Basic Books.
Lynch, K. 1962. *The image of the city*. Cambridge, Mass.: MIT Press.
Lyotard, J.-F. 1984. *Tombeau de l'intellectuel et autres papiers*. Paris: Galilée.

McClagan, D. 1977. *Creation myths: Man's introduction to the world*. Singapore: Thames & Hudson.
Mackenzie, S., ed. 1986. *Humanism and geography*. Carleton University, Department of Geography Discussion Papers. Ottawa.
MacLaughlin, J. 1986. State centered social science and the anarchist critique: Ideology in political geography. *Antipode* 18: 11–38.
Mahrt, W. P. 1979. Gregorian chant as a fundamentum of Western musical culture: An introduction to the singing of a solemn high mass. *Bulletin of the American Academy of Arts and Sciences* 3: 22–34.
Mair, A. 1986. Thomas Kuhn and understanding geography. *Progress in Human Geography* 10: 345–70.
Malinowski, B. 1955. *Magic, science, and religion*. New York: Doubleday.
Mandelbrot, B. B. 1983. *The fractal geometry of nature*. New York: Freeman.
Mannheim, K. 1946. *Ideology and utopia: An introduction to the sociology of knowledge*. New York: Harcourt, Brace, & World.
March, T., and G. Overwold, eds. 1985. *Interpreting the humanities*. Princeton, N.J.: Woodrow Wilson National Fellowship Foundation.
Marchand, B. 1982. Dialectical analysis of value: The example of Los Angeles. In *A search for common ground*, ed. P. Gould and G. Olsson, 232–51. London: Pion.
Marcuse, H. 1972. *Studies in critical philosophy*. Boston: Beacon Press.
Marlowe, C. [1587] 1967. *Tamburlaine the Great*. Ed. John D. Jump. London: Edward Arnold.
———. [1604] 1616. 1982. *Doctor Faustus*. In C. W. Eliot, ed., *Johann Wolgang von Goethe and Christopher Marlowe*, 203–50. Harvard Classics. Danbury, Conn.: Grolier.
Marsh, G. P. 1864. *Man and nature, or physical geography as modified by human action*. New York: Scribner.
Mårtensson, S. 1979. *On the formation of biographies in space-time environments*. Lund Studies in Human Geography, ser. B, no. 47. University of Lund.
Matos, L. 1960. La littérature des découvertes. In *Les aspects intérnationaux de la découverte océanique au XVe et XVIe siècles*. Colloque international d'histoire maritime, 5: 23–30. Paris: Ecole Pratique des Hautes Etudes.
Matthiessen, P. [1979] 1987. *The snow leopard*. Harmondsworth, Middlesex: Penguin Books.
May, J. A. 1970. *Kant's conception of geography and its relation to recent geographical thought*. University of Toronto, Department of Geography Research Publications.
Mead, W. R. 1981. *An historical geography of Scandinavia*. London: Academic Press.
Meinig, D. W. 1976. The beholding eye: Ten versions of the same scene. *Landscape Architecture* (January): 47–64.
———, ed. 1979. *The interpretation of ordinary landscapes: Geographical essays*. New York: Oxford University Press.

Mendelsohn, E., P. Weingart, and R. Whitley, eds. 1977. *The social production of scientific knowledge: Sociology of sciences yearbook.* Dordrecht, Netherlands: Reidel.

Merleau-Ponty, M. [1947] 1969. *Humanism and terror.* Trans. J. O'Neill. Boston: Beacon Press.

Michelet, J. 1833–67. *Histoire de France.* 17 vols. Paris: Hachette.

Midgley, M. 1989. *Wisdom, information, and wonder: What is knowledge for?* London: Routledge.

Mikesell, M. W. 1968. Friedrich Ratzel. *International Encyclopedia of the Social Sciences,* 13: 327–29.

Mills, W. J. 1982. Metaphorical vision: Changes in Western attitudes toward the environment. *Annals of the Association of American Geographers* 72: 237–53.

Monod, J. 1972. *Chance and necessity.* New York: Vintage Books.

Montaigne, M., Seigneur de. [1603] 1958. *Essais.* Trans. J. Florio as *Essays.* Harmondsworth, Middlesex: Penguin Books.

Montesquieu, C. de Secondat Baron de la Brède et de Montesquieu. [1853] 1950–51. Défense de l'esprit des lois. In *Oeuvres complètes de Montesquieu,* ed. André Masson. 3 vols. Paris: Editions Firmin-Didot.

Moore, A. 1957. *The frontier mind: A cultural analysis of the Kentucky frontiersman.* Lexington: University of Kentucky Press.

Moore, G., and R. Golledge, eds. 1976. *Environmental knowing: Theories, research, and methods.* Stroudsburgh, Pa.: Hutchinson & Ross.

Morgan, G. 1980. Paradigms, metaphors, and puzzle-solving in organization theory. *Administrative Science Quarterly* 25: 605–22.

Morin, E. 1973. *Le paradigme perdu: La nature humaine.* Paris: Editions du Seuil.

Müller, M. 1871. Metaphor. In *Lectures in the science of language,* ed. M. Müller, 351–402. 2d ser. New York: Scribner.

Mumford, L. [1934] 1962. *Technics and civilization.* Reprint. New York: Harcourt, Brace.

Mungall, C., and D. J. McLaren. 1990. *Planet under stress: The challenge of global change.* Toronto: Oxford University Press.

Murdie, R. A. 1969. *The factorial ecology of metropolitan Toronto.* Chicago: University of Chicago Press.

Murray, K., and C. Murray. 1980. *Illuminations from the Bhagavad-Gita.* New York: Harper Colophon.

Myers, D. 1982. From the duck pond to the global commons: Increasing awareness of the supranational nature of emerging environmental issues. *Ambio* 11: 195–201.

Nakamura, H. 1980. The idea of nature, east and west. In *The great ideas today,* 234–304. Chicago: Encyclopedia Britannica.

Nash, P. 1985. Becoming a humanist geographer: A circuitous journey. In Guelke, 1–22.

Needham, J. 1965. *Time and Eastern man.* Occasional Paper no. 21. Royal Anthropological Institute, London.

Neihardt, J. G. 1961. *Black Elk speaks: Being the life story of a holy man of the Oglala Sioux.* Lincoln: University of Nebraska Press.

Nelson, H. 1913. Hembygdsundervisning i folkhögskolan. *Svenska folkhögskolans årsbok,* 25–35.

Newman, Sir Henry Cardinal. 1852. *The idea of a university defined and illustrated.* London: Longmans.
Newton, I. [1687] 1729. *Philosophiae naturalis principia mathematica.* Trans. A. Motte as *The mathematical principle of natural philosophy*. London: Benjamin Motte.
Nicolini, F., and B. Croce, eds. 1911–41. *Opere di G. B. Vico.* 8 vols. Bari: Laterza.
Nicolson, M. 1960. *The breaking of the circle: Studies in the effect of the "new science" upon seventeenth century poetry.* New York: Oxford University Press.
Nisbet, R. 1953. *The quest for community.* New York: Oxford University Press.
Norberg-Schulz, C. 1980. *Genius loci: Toward a phenomenology of architecture.* New York: Rizzoli.

Oates, W. J., and E. O'Neill, eds. 1938. *Seven famous Greek plays.* New York: Vintage Books.
Olsson, G. 1979. Social science and human action; or, on hitting your head against the ceiling of language. In *Philosophy in geography,* ed. S. Gale and G. Olsson, 287–308. Dordrecht, Netherlands: Reidel.
———. 1984. Toward a sermon of modernity. In Billinge et al., 73–85.
Olwig, K. 1984. *Nature's ideological landscape.* London: Allen & Unwin.
O'Neill, T. 1982. *Essaying Montaigne: A study of the Renaissance institution of writing and reading.* London: Routledge & Kegan Paul.
Ortega y Gasset, J. 1957. *Man and people.* New York: Norton.
Ortony, A. 1979. *Metaphor and thought.* Cambridge: Cambridge University Press.

Padovano, A. T. 1975. Journey. In *Journeys,* ed. G. Baum, 210–35. New York: Paulist Press.
Palmer, R. E. 1969. *Hermeneutics.* Evanston, Ill.: Northwestern University Press.
Park, R. E., E. Burgess, and R. McKenzie. 1925. *The City.* Chicago: University of Chicago Press.
Pascal, B. 1670. *Pensées.* Paris: Guillaume Deprez.
Pascon, P. 1979. De l'eau du ciel à l'eau de l'Etat: Psycho sociologie de l'irrigation au Maroc. *Hérodote* 13: 60–78.
Passet, R. 1979. *L'économique et le vivant.* Paris: Payot.
Passmore, J. 1974. *Man's responsibility for nature: Ecological problems and Western traditions.* New York: Scribner.
Peet, R. 1977. *Radical geography: Alternative viewpoints on contemporary social issues.* Chicago: Maroufa Press.
Peirce, C. S. 1931–35. *Collected papers.* Ed. C. Hartshorne and P. Weiss. Cambridge, Mass.: Harvard University Press.
Pelt, J.-M. 1977. *L'homme re-naturé: Vers la société écologique.* Paris: Editions du Seuil.
Pepper, D. 1984. *The roots of modern environmentalism.* London: Croom Helm.
Pepper, S. C. 1942. *World hypotheses.* Berkeley: University of California Press.
Perroux, F. 1954. *L'Europe sans rivages.* Paris: Presses universitaires de France.
Phillips, D. C. 1970. Organicism in the late nineteenth and early twentieth centuries. *Journal of the History of Ideas* 31: 413–32.
Phipps, M. 1981a. Entropy and community pattern analysis. *Journal of Theoretical Biology* 93: 253–73.
———. 1981b. Information theory and landscape analysis. In *Perspectives in landscape*

ecology, ed. S. P. Tjallingii and A. A. de Veer, 57–64. Wageningen, Netherlands: Pudoc.

Phipps, M., and V. Berdoulay. 1985. *Paysage et système.* Ottawa: Editions de l'Université d'Ottawa.

Phipps, M., et al. 1986. Ordre topo-écologique dans un paysage rural: Les niches paysagiques. *Comptes rendus des sciences de Paris,* vol. 302, ser. 3, no. 20, pp. 691–96.

Plato. 1952. *Timaeus, Critias, Cleitophon, Menexenus, Epistolae.* Trans. R. G. Bury. Loeb Classical Library. Cambridge, Mass.: Harvard University Press.

Pocock, D.C.D., ed. 1981. *Humanistic geography and literature: Essays on the experience of place.* London: Croom Helm.

———. 1988. Geography and literature. *Progress in Human Geography* 12: 87–102.

Poincaré, H. [1914] 1955. *Science and Method.* New York: Dover.

Polanyi, M. 1958. *Personal knowledge: Toward a post-critical philosophy.* Chicago: University of Chicago Press.

Pope, A. [1733–34] 1969. *Essay on man.* Facs. reproduction. London: Menston.

Popper, K. 1976. *Unended quest: An intellectual autobiography.* La Salle, Ill.: Open Court.

Porteous, J. D. 1990. *Landscapes of the mind: Worlds of sense and metaphor.* Toronto: University of Toronto Press.

Powell, J. W. 1878. A discourse on the philosophy of the North American Indians. *Journal of the American Geographical Society* 8: 251–68.

Pred, A. 1977. *City systems in advanced economies: Past growth, present processes, and future development options.* London: Hutchinson.

———. 1984. Place as historically contingent process: Structuration and the time-geography of becoming places. *Annals of the Association of American Geographers* 74 (2): 279–98.

Prigogine, I., and I. Stengers. 1979. *La nouvelle alliance: Métamorphose de la science.* Paris: Gallimard.

———. 1984. *Order out of chaos: Man's new dialogue with nature.* Boulder: Shambhala, New Science Library.

Racine, J.-B. 1977. Discours géographique et discours idéologique: Perspectives épistemologiques et critiques. *Hérodote* 6: 109–58.

Racine, J.-B., and C. Raffestin. 1983. L'espace et la société dans la géographie sociale francophone: Pour une approche critique du quotidien. In *Espace et localisation, la redécouverte de l'espace dans la pensée scientifique de langue français,* ed. J.H.P. Paelinck and A. Sallez, 304–30. Paris: Economica.

Ratzel, F. 1882–91. *Anthropogeographie.* 2 vols. Vol. 1, *Grundzüge der anwendung der Erdkunde auf die Geschichte;* vol. 2, *Die geographische Verbreitung des Menschen.* Stuttgart: Engelhorn.

———. [1897] 1923. *Politische Geographie.* 3d ed. Ed. E. Oberhummer. Munich: Oldenbourg.

Ray, J. [1690] 1759. *The Wisdom of God manifested in the works of the creation.* 12th ed. London: John Rivington, John Ward, Joseph Richardson.

Reclus, E. 1877. *Nouvelle géographie universelle: La terre et les hommes.* Paris: Hachette.

Reed, A. W. 1980. *Aboriginal stories of Australia.* New South Wales: Reed Books.

Reichenbach, H. [1951] 1973. *The rise of scientific philosophy.* Berkeley: University of California Press.

Relph, E. 1974. *Place and placelessness.* London: Pion.

———. 1981. *Rational landscapes and humanistic geography.* London: Croom Helm.

Richardson, M. 1982. Being-in-the-market versus being-in-the-plaza: Material culture and the construction of social reality in Spanish America. *American Ethnologist* 9: 421–36.

Richter, J. P., ed. 1970. *The literary works of Leonardo da Vinci.* London: Phaidon.

Ricoeur, P. 1971. The model of the text: Meaningful action considered as a text. *Social Research* 38: 529–62.

———. 1975. *La métaphore vive.* Paris: Editions du Seuil.

Ritter, C. 1817. *Allgemeine Erdkunde: Vorlesungen an der Universität zu Berlin* (*Comparative geography*). Edinburgh: W. Blackwood.

Rocchi, J. 1989. *L'errance et l'hérésie ou le destin de Giordano Bruno.* Paris: Editions François Bourin.

Rochefort, R. 1961. *Le travail en Sicile: Etude de géographie sociale.* Presses universitaires de France.

Rorty, R. 1979. *Philosophy and the mirror of nature.* Princeton, N.J.: Princeton University Press.

———. 1982. Hermeneutics, general studies, and teaching. *Synergos: Selected papers from the Synergos seminars,* no 2. Fairfax, Va.: George Mason University Press.

Rose, C. 1981. Wilhelm Dilthey's philosophy of historical understanding: A neglected heritage of contemporary humanistic geography. In Stoddart 1981, 99–133.

———. 1987. The problem of reference and geographic structuration. *Environment and planning,* D.: *Society and Space,* 5: 93–106.

Rostlund, E. 1962. Twentieth century magic. In *Readings in cultural geography,* ed. P. Wagner and M. Mikesell, 48–54. Chicago: University of Chicago Press.

Rowles, G. D. 1978. *Prisoners of space? Exploring the geographical experience of older people.* Boulder, Colo.: Westview Press.

Rowntree, L. 1986, 1988. Cultural/humanistic geography. *Progress in human geography* 10: 580–86; 12: 575–86.

Rühl, A. 1938. *Einfuhrung in die allgemeine Wirtsacharftsgeographie.* Leiden: A. W. Suthoff's Uitgeversmaatschaapij, N.V.

Russell, B. 1959. *Wisdom of the West.* London: Rathbone Books.

Saarinen, T., D. Seamon, and J. L. Sell, eds. 1984. *Environmental perception and behavior: Inventory and prospect.* University of Chicago, Department of Geography Research Paper no. 209.

Sachs, S., ed. 1979. *On metaphor.* Chicago: University of Chicago Press.

Sack, D. 1991. The trouble with antithesis: The case of G. K. Gilbert, geographer and educator. *Professional Geographer* 43 (1): 28–37.

Sack, R. 1980a. Concepts of geographic space. *Progress in Human Geography* 4 (3): 315–45.

———. 1980b. *Conceptions of space in social thought.* London: Macmillan.

Sadat, A. L. 1977. *In search of identity: An autobiography.* New York: Harper & Row.

Said, E. W. 1983. Opponents, audiences, constituencies, and community. In *Anti-*

aesthetic: Essays on postmodern culture, ed. H. Foster, 135–59. Port Townsend, Wash.: Bay Press.

Sällstrom, P. 1986. *Mentaliteter.* Åbo, Finland: Publications of the Research Institute of the Åbo Akademi Foundation.

Salter, C. L. 1978. Signatures and settings: An approach to landscape in literature. In Butzer 1978, 69–83.

Sambursky, S. [1956] 1987. *The physical world of the Greeks.* Princeton, N.J.: Princeton University Press.

Samuels, M. 1971. Science and geography: An existential appraisal. Ph.D. diss., Department of Geography, University of Washington, Seattle.

Santos, M. [1975] 1985. *L'espace partage.* Paris: Editions M. Th. Genin. Adapted for publication in English by Chris Gerry as *The shared space.* New York: Methuen.

Sauer, C. O. 1925. The morphology of landscape. In *University of California publications in geography,* 2: 19–53. Department of Geography.

———. 1941. Foreword to historical geography. *Annals of the Association of American Geographers* 31: 1–24.

Sayer, A. 1982. Explanation in economic geography: Abstraction versus generalization. *Progress in Human Geography* 6: 68–88.

Schlanger, A. J. 1971. *Les métaphores de l'organisme.* Paris: Vrin.

Schoder, R. V. 1974. *Ancient Greece from the air.* London: Thames & Hudson.

Schofield, R. E. 1970. *Mechanism and materialism: British natural philosophy in an age of reason.* Princeton, N.J.: Princeton University Press.

Schouw, J. F. 1925. Selfbiografi. *Botanisk tidskrift* 38: 3–4.

Schrag, C. O. 1980. *Radical reflection and the origin of the human sciences.* West Lafayette, Ind.: Purdue University Press.

Schrödinger, E. 1967. *What is life? Mind and matter.* Cambridge: Cambridge University Press.

Schultz, H.-D. 1980. *Die deutschsprachige Geographie von 1800 bis 1970: Ein bitrag zur Geschichte ihrer Methodologie.* Berlin: Selbstverlag des Geographischen Instituts der Freien Universität Berlin.

Schütz, A. 1944. The stranger: An essay in social psychology. *American Journal of Sociology* 49: 499–507.

———. 1962. *Collected papers.* 2 vols. The Hague: Martinus Nijhoff.

———. 1973. *Structures of the lifeworld.* Ed. T. Luckman. Evanston, Ill.: Northwestern University Press.

Schwenk, T. 1976. *Sensitive chaos.* New York: Schocken Books.

Scott, A. 1982. The meaning and spatial origins of discourse on the spatial foundations of society. In Gould and Olsson, 141–56.

Seamon, D. 1980. *A geography of the lifeworld: Movement, rest, and encounter.* London: Croom Helm.

Seamon, D., and R. Mugerauer, eds. 1985. *Dwelling, place, and environment.* Dordrecht, Netherlands: Martinus Nijhoff.

Semple, E. C. 1911. *Influences of geographical environment on the basis of Ratzel's system of anthropogeography.* New York: Holt.

———. 1931. *The geography of the Mediterranean region: Its relation to ancient history.* New York: Holt.

Shaler, N. 1894. *The United States of America.* 3 vols. New York: Appleton.

———. 1904. *The neighbor: The natural history of human contacts.* New York: A. P. Watt.

Sheldrake, R. 1981. *A new science of life.* London: Blond & Briggs.

Shelley, P. B. [1820] 1881. *Prometheus unbound: A lyrical drama in four acts.* Ed. K. Raine. Harmondsworth, Middlesex: Penguin Books.

Shen, Shurong. 1985. *Water symbols in ancient Chinese: Hydrological and geological history* (in Chinese). Beijing: Geological Publishing.

Shibles, W. A. 1971. *Metaphor: An annotated bibliography and history.* Whitewater, Wisc.: Language Press.

Singh, R. L., and R.P.B. Singh. 1984. Lifeworld and lifecycle in India: A search in geographical understanding. *National Geographical Journal of India* 30: 207–22.

Sion, J. 1909. *Les paysans de la Normandie orientale: Pays de Caux, Bray, Véxin Normand, Vallée de la Seine: Etude géographique.* Paris: Armand Colin.

———. 1934. *Méditerranée: Péninsules méditerranéennes.* Paris: Armand Colin.

Smith, J. R. 1907. Economic geography and its relation to economic theory and higher education. *Bulletin of the American Geographical Society* 39: 8.

———. 1925. *North America.* New York: Harcourt, Brace.

Smith, N. 1979. Geography, science, and post-positivist modes of explanation. *Progress in Human Geography* 3: 356–83.

Smith, S. 1984. Practicing humanistic geography. *Annals of the Association of American Geographers* 74: 353–74.

Söderquist, T. 1986. *The ecologists: From merry naturalists to saviours of the nation.* Stockholm: Almqvist & Wiksell International.

Solot, M. 1986. Carl Sauer and cultural evolution. *Annals of the Association of American Geographers* 76: 508–20.

Sophocles. 1948. *The three Theban plays.* Trans. R. Fagles. Harmondsworth, Middlesex: Penguin Classics.

Sorre, M. 1943. *Les bases biologiques de la géographie humaine: Essai d'une écologie de l'homme.* Paris: Armand Colin.

———. 1961. *L'homme sur la terre: Traité de géographie humaine.* Paris: Hachette.

Sörlin, S. 1988. *Debatten om Norrland och naturresurserna under det industriella genombrott.* Stockholm: Carlssons. (English summary, 262–72.)

Spate, O.H.K. [1960] 1966. Quantity and quality in geography. In Spate, *Let me enjoy: Essays partly geographical,* 149–82. London: Methuen.

Spea, J. J. 1948. *Itó jensai.* Peking: Catholic University of Peking.

Speth, W. 1981. Berkeley geography. In *The origins of academic geography in the United States,* ed. B. Blouet, 221–44. Hamden, Conn.: Archon Books.

Spiegelberg, F. 1961. *Zen, rocks and waters.* New York: Random House.

Steiner, G. 1987. Some black holes. *Bulletin of the American Academy of Arts and Sciences* 51: 12–28.

Sterritt, N. 1991. Land and life: Gitskan experiences. In *Land life leisure lumber: Local and global concern in the human use of woodland,* ed. A. Buttimer, J. van Buren, and N. Hudson-Rodd, 94–112. Ottawa: University of Ottawa Department of Geography and Royal Society of Canada.

Stewart, J. Q. 1953. Social physics and the constitution of the United States. Address to the Foundation for Integrated Education, April 29, Philadelphia.

Stoddart, D. R. 1966. Darwin's impact on geography. *Annals of the Association of American Geographers* 56: 683–98.

———. 1967. Organism and eco-systems as geographical models. In Chorley and Haggett, 511–48.

———, ed. 1981. *Geography, ideology, and social concern*. Oxford: Basil Blackwell.

Stokes, W. L. 1973. *Essentials of earth history: An introduction to historical geology.* Englewood Cliffs, N.J.: Prentice-Hall.

Stone, K. H. 1979. Geography's wartime service. *Annals of the Association of American Geographers* 69: 89–96.

Strahler, A. N. 1952. The dynamic basis of geomorphology. *Bulletin of the Geological Society of America* 63: 923–37.

Sugiura, N. 1983. Rhetoric and geographers' worlds: The case of spatial analysis in human geography. Ph.D. diss., Department of Geography, Pennsylvania State University, State College.

Tagliacozzo, G., M. Mooney, and D. P. Verene, eds. 1969. *Vico and contemporary thought*. 2 vols. Atlantic Highlands, N.J.: Humanities Press.

Taylor, G., ed. 1951. *Geography in the twentieth century.* London: Methuen.

———. 1958. *Journeyman Taylor: The education of a scientist*. London: Robert Hale.

Teich, M., and R. Young, eds. 1973. *Changing perspectives in the history of science.* London: Heinemann.

Teilhard de Chardin, P. [1955] 1959. *Le phenomène humain*. Paris: Editions du Seuil. Trans. B. Wall as *The phenomenon of man*. London: W. Collins; New York: Harper & Row.

———. [1959] 1964. *L'avenir de l'homme*. Paris: Editions du Seuil. Trans. N. Denny as *The future of man*. London: W. Collins; New York: Harper & Row.

Theodorson, G. A. 1958. *Studies in human ecology.* Evanston, Ill.: Row, Petersen.

Thom, R. 1975. *Structural stability and morphogenesis*. New York: Benjamin.

Thomas, W. L., ed. 1956. *Man's role in changing the face of the earth*. Chicago: University of Chicago Press.

Thomas Aquinas, Saint. [1244] 1949. *De Regno, Ad Regem Cypri*. Trans. G. B. Phelan as *On kingship to the king of Cyprus*. Toronto: Pontifical Institute of Medieval Studies.

———. [1253] 1965. *On being and essence*. Trans. Joseph Bobik. Notre Dame, Ind.: University of Notre Dame Press.

Thompson, D'Arcy W. 1917. *On growth and form*. Cambridge: Cambridge University Press.

Thompson, J. P. 1859. The value of geography to the scholar, the merchant, and the philanthropist. *Journal of the American Geographical and Statistical Society* 1: 98–107.

Thorndike, L. 1955. The true place of astrology in the history of science. *Isis* 46: 273–78.

Toulmin, S. 1983. *The return to cosmology.* Berkeley: University of California Press.

Tuan, Yi-Fu. 1968. *The hydrological cycle and the wisdom of God: A theme in geoteleology.* Toronto: University of Toronto Press and Department of Geography Research Publications, no. 1.

———. 1971. Geography, phenomenology, and the study of human nature. *Canadian Geographer* 15: 181–92.

———. 1976. Humanistic geography. *Annals of the Association of American Geographers* 66: 266–76.

———. 1978. Sign and metaphor. *Annals of the Association of American Geographers* 68: 363–72.
———. 1982. *Segmented worlds and the self: Group life and individual consciousness.* Minneapolis: University of Minnesota Press.
Turner, F. 1980. *Beyond geography: The Western spirit against the wilderness.* New York: Viking Press.
Turner, F. J. [1893] 1961. The significance of the frontier in American history. In Turner, *Frontier and section: Selected essays,* 37–62. Englewood Cliffs, N.J.: Prentice-Hall.
———. 1920. *The frontier in American history.* New York: Holt.
———. 1961. *Frontier and section: Selected essays of Frederick Jackson Turner.* Englewood Cliffs, N.J.: Prentice-Hall.

Ullman, E. A. 1940–41. A theory of location for cities. *American Journal of Sociology* 46: 853–64.
———. 1954. Geography as spatial interaction. *Annals of the Association of American Geographers* 68: 363–72.
Urry, J. 1981. Localities, regions, and class. *International Journal of Urban and Regional Research* 5: 455–73.

Vallaux, C. 1925. *Les sciences géographiques.* Paris: Alcan.
van Fraassen, B. 1980. *The scientific image.* Oxford: Oxford University Press.
Van Valkenburg, S. 1951. The German school of geography. In G. Taylor 1951, 91–115.
Varenius. 1672. *Geographia Generalis, in qua affectiones generales Telluris explicantur.* Ed. I. Newton. Cambridge: Cambridge University Press.
Vico, G. B. [N.d.] 1944. *The autobiography of Giambattista Vico.* Trans. M. H. Fish and T. B. Bergin. Ithaca, N.Y.: Cornell University Press.
———. [1744] 1948. *The new science of Giambattista Vico.* Trans. T. B. Bergin and M. H. Fish. Ithaca, N.Y.: Cornell University Press.
Vidal de la Blache, P.-M. 1896. La géographie politique: A propos des écrits de M. Frédéric Ratzel. *Annales de géographie* 7: 97–111.
———. 1903. *Tableau de la géographie de la France.* Paris: Hachette.
———. 1913. Les caractères distinctifs de la géographie. *Annales de géographie* 22: 289–99.
———. 1917. *La France de l'Est.* Paris: Armand Colin.
———. [1922] 1926. *Principes de géographie humaine.* Ed. E. de Martonne. Paris: Armand Colin. Trans. M. T. Bingham as *Principles of human geography.* London: Constable.
Voget, F. W. 1975. *A history of ethnology.* New York: Holt, Rinehart, & Winston.
Voltaire. [1759] 1959. *Candide.* Trans. L. Bair. London: Bantam Books.
Von Wright, G. H. 1978. *Humanismen som livshållning ocg andra essayer.* Stockholm: Rabén & Sjögren.

Warntz, W. 1964. *Geography now and then.* Research Series, no. 25. New York: American Geographical Society.
———. 1984. Trajectories and co-ordinates. In Billinge et al., 134–52.

Weber, M. [1930] 1958. *The Protestant ethic and the rise of capitalism.* Trans. T. Parsons. New York: Scribner; London: Allen & Unwin.

Weiner, D. R. 1988. *Models of nature, ecology, conservation, and cultural revolution in Soviet Russia.* Bloomington: Indiana University Press.

Western, J. 1986. Places, authorship, authority: Retrospection on fieldwork. In Guelke, 23–38.

Whicher, S. E. 1953. *Freedom and fate: An inner life of Ralph Waldo Emerson.* Philadelphia: University of Pennsylvania Press.

White, G. F., ed. 1974. *Natural hazards: Local, national, global.* New York: Oxford University Press.

———. 1985. Geographers in a perilously changing world. *Annals of the Association of American Geographers* 75, no. 1: 1–10.

White, H. 1974. *Metahistory: The historical imagination in nineteenth-century Europe.* Baltimore: Johns Hopkins University Press.

White, L. 1967. The historical roots of our ecological crisis. *Science* 155: 1203–7.

Whitehead, A. N. [1933] 1969. *Adventures of ideas.* New York: Macmillan.

Whittlesey, D. W. 1929. Sequent occupance. *Annals of the Association of American Geographers* 19: 162–65.

Wiener, P. 1949. *Evolution and the founders of pragmatism.* Cambridge, Mass.: Harvard University Press.

William-Olsson, W. 1937. *Huvuddrag av Stockholms geografiska utveckling, 1850–1930.* Pt. 2, *Nutida Stockholm med förorter.* Stadskollegiets Utlåtanden och Memorial. Stockholm: Akademisk Avhandling.

Wilson, A. G. 1970. *Entropy in urban and regional modelling.* Centre for Environmental Studies Working Paper no. 26. London.

Wittgenstein, L. 1969. *On certainty.* Oxford: Basil Blackwell.

WCED (World Commission on Environment and Development). 1987. *Our common future.* London: Oxford University Press.

Wolch, J., and M. Dear, eds. 1988. *The power of geography: How territory shapes social life.* London: Unwin Hyman.

Wooldridge, S. W., and W. G. East. 1952. *The spirit and purpose of geography.* London: Hutchinson's University Library.

Worster, D. 1986. *Rivers of empire: Aridity and growth of the American West.* New York: Pantheon Books.

Wright, J. K. 1925. *The geographical lore of the time of the Crusades: A study in the history of medieval science and tradition in Western Europe.* New York: American Geographical Society.

———. 1952. *Geography in the making: The American Geographical Society, 1851–1951.* New York: American Geographical Society.

———. 1961. Daniel Coit Gilman: Geographer and historian. *Geographical Review* 51: 381–99.

———. 1966. *Human nature in geography: Fourteen papers, 1925–65.* Cambridge, Mass.: Harvard University Press.

Zaehner, R. C., ed. 1966. *Hindu scriptures.* London: J. M. Dent.

Zelinsky, W. 1970. Beyond the exponentials. *Economic Geography* 46: 498–535.

Zimmermann, C. C., and M. E. Frampton, 1935. *Family and society: A study of the sociology of reconstruction.* New York: Holt.
Zonneveld, J.I.S. 1984. Geography and poetry. In *Modern geographic trends,* ed. R. L. Singh and R.P.B. Singh, 13–40. Felic. vol. for Prof. E. Ahmad. Ranchi, India: N.p.

Index

Index

Index

Index

Index

Index

Index

About the Author

Anne Buttimer is professor and head of the Geography Department at University College Dublin. She received her Ph.D. in geography at the University of Washington in Seattle. She has held research and teaching positions in Belgium, France, Scotland, and Sweden and has been professor of geography at Clark University and the University of Ottawa. Her articles have appeared in leading professional journals, including *Annals of the Association of American Geographers, Geographical Review, L'Espace Géographique, Münchener Geographische Hefte, Geografiska Annaler, Pacific Viewpoint,* and *Town Planning Review*. Some of her work has been published in translation to Dutch, French, German, Japanese, Portuguese, Russian, Spanish, and Swedish. She is author of *The Wake of Erasmus* (1989), *The Practice of Geography* (1983), *Values in Geography* (1974), and *Society and Milieu in the French Geographic Tradition* (1971); and co-author of *Geographers of Norden* (1988), *Creativity and Context* (1983), and *The Human Experience of Place and Space* (1980). She has received over a dozen awards and honors, among them the Ellen Churchill Semple Award (1991) and the Association of American Geographers' Honors Award (1986). She has been secretary of the International Geographical Union Commission on the History of Geographical Thought since 1988 and directed a Swedish-Canadian research exchange on the human experience of woodland (1989–92). She is currently coordinating an international research project on transformations of landscape and life with partner teams in Germany, Ireland, the Netherlands, and Sweden.